DA PAMPHLET 750-33
DECEMBER 1976

CHARGING SYSTEM TROUBLESHOOTING

(The Easy Way)

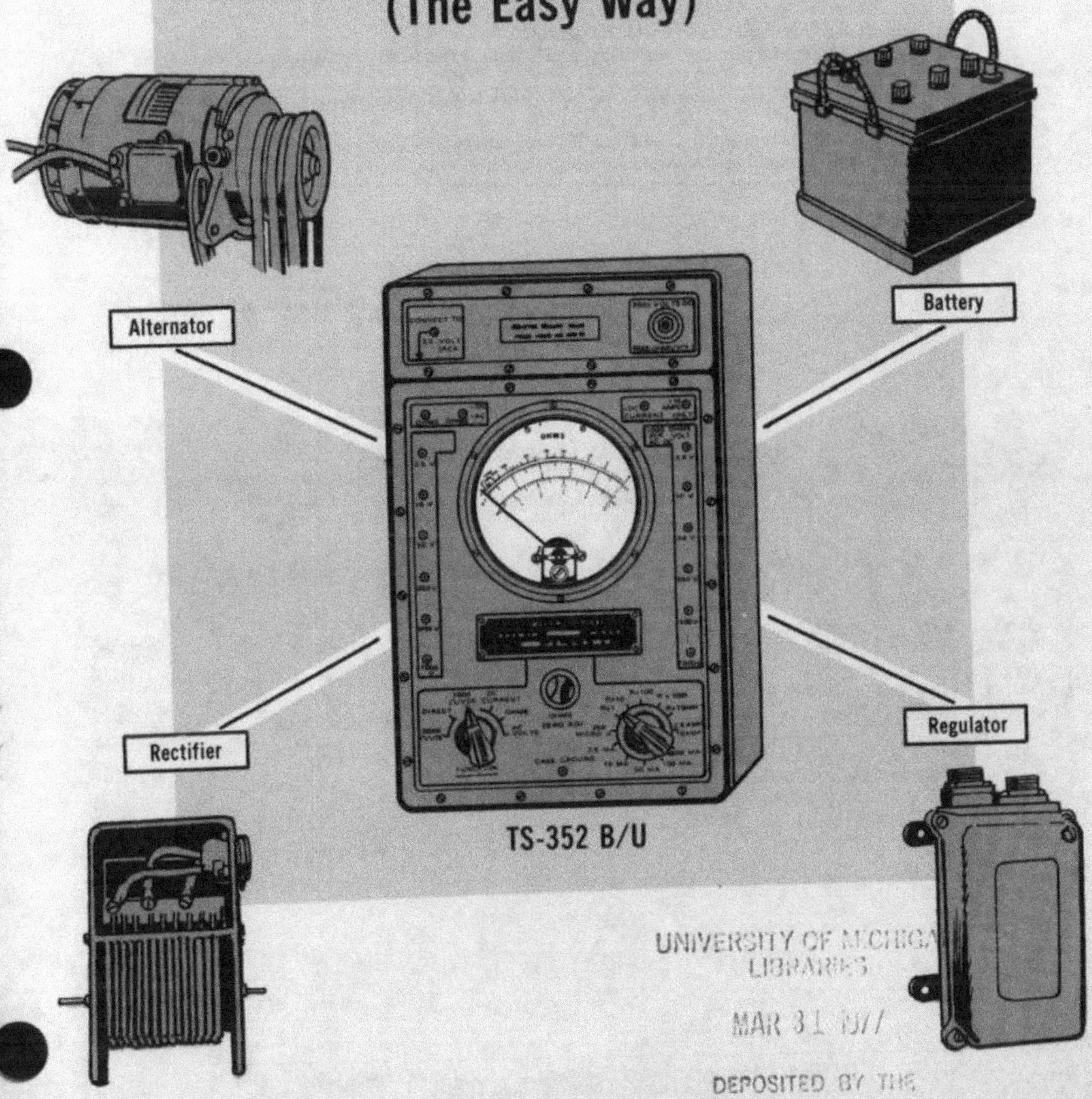

Alternator

Battery

Rectifier

Regulator

TS-352 B/U

WHAT THIS PAMPHLET'S ALL ABOUT . . . .

The electricity your vehicle needs comes from the batteries and the alternator/generator. That same electricity powers the starter, fires the sparkplugs, keeps the lights burning and operates communications and weapons systems.

Your batteries provide the initial power, and then the alternator/generator takes over . . . recharging the batteries and providing power for other purposes.

Your cables and connections and your regulator are the final link in your vehicle's charging system (your regulator keeps the generator from putting out too much or too little electricity).

When any part of the "system" breaks down, your vehicle gets too much or too little electrical power, like so:

- Dead or discharged batteries can't crank the starter.
- Faulty generators or regulators can overcharge and boil the water from your batteries. Overcharging also burns out your lights.
- Commo and weapons systems won't work right when the charging system puts out too little electricity.

Use of a few basic tools, a multimeter and an antifreeze/battery tester, as described in this pamphlet, pinpoints your charging system problem and allows you to correct it . . . quickly and with a minimum of sweat. Use of the basic tools and test equipment is fully described in the text of this pamphlet.

DA Pamphlet 750-33
Headquarters, Department of the Army
Washington, D.C.
15 December 1976

CHARGING SYSTEM TROUBLESHOOTING

Table of Contents

		Page
INTRODUCTION		Inside Front Cover
Chapter 1.	Use of multimeters, Antifreeze/Battery Tester and tools list	1
Chapter 2.	Battery Voltage Checks	8
Chapter 3.	25-Amp Charging System Tests (wheeled veh.)	10
Chapter 4.	60-AMP Charging System Tests (wheeled veh.)	14
Chapter 5.	100-AMP Internally Rectified (wheeled veh.)	18
Chapter 6.	100-AMP Externally Rectified (wheeled veh.)	23
Chapter 7.	100-AMP Internally Rectified (Track)	29
Chapter 8.	100-AMP Externally Rectified, Plate (Track)	34
Chapter 9.	100-AMP Externally Rectified, Molded (Track)	39
Chapter 10.	300-AMP (M60-series, M88)	45
Chapter 11.	300-AMP (M107, M110, M578)	52
Chapter 12.	300-AMP (M551-series)	59

For sale by the Superintendent of Documents, U.S. Government Printing Office
Washington, D.C. 20402 - Price $1.40

Stock No. 008-020-00623-8

Chapter 1
GENERAL INSTRUCTIONS FOR USE OF MULTIMETERS

In automotive troubleshooting, the TS-352B/U, and AN/URM-105 and the Simpson 160 will do the same job. Therefore, your automotive shop sets may contain any one of these multimeters.

(Fig 1-1)

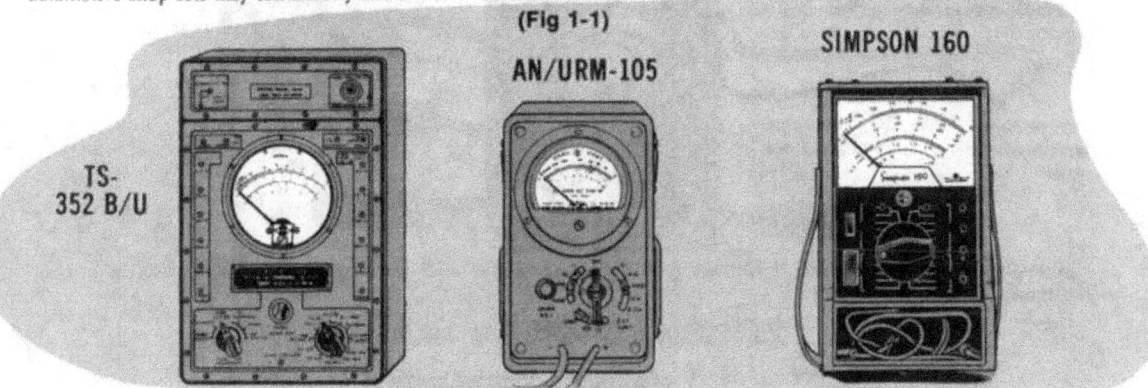

AN/URM-105

SIMPSON 160

TS-352 B/U

Any of these multimeters can be used to troubleshoot your vehicle's electrical system. This section shows how.

MECHANICAL METER ADJUSTMENT

Instruments that have a meter movement read-out and a mechanical adjustment, such as the multimeters TS-352, AN/URM-105 and the Simpson 160, may require adjusting prior to their use.

(Fig 1-2)

Needle should be adjusted so that with one eye closed and looking down at needle it cuts the zero in half on the left side of the scale.

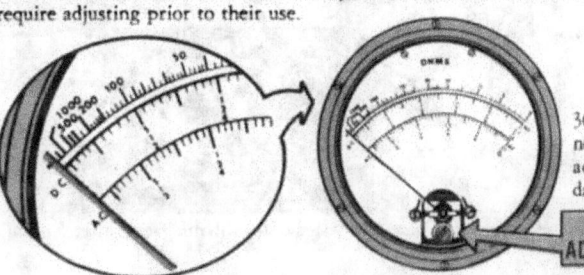

(Fig 1-3)

Turning this screw 360° will only cause the needle to move. This adjustment will not damage the meter.

MECHANICAL ADJUSTING SCREW

If the needle cannot be adjusted to zero by turning the screw clockwise or counter-clockwise, the instrument is faulty and needs repair.

If the needle zeros when the instrument is upright and is off when it is laid down, readjust the zero setting for the position in which the instrument will be used.

ELECTRICAL METER ADJUSTMENT

The multimeter must be set up and "zeroed" before making tests in this pamphlet. The following steps on each meter are the ones you do to set the meter up to read ohms, continuity and shorts. Therefore, it is necessary to make sure that the meter will "zero" before using it. Match the following steps to the meter you have and do them to electrically zero the meter:

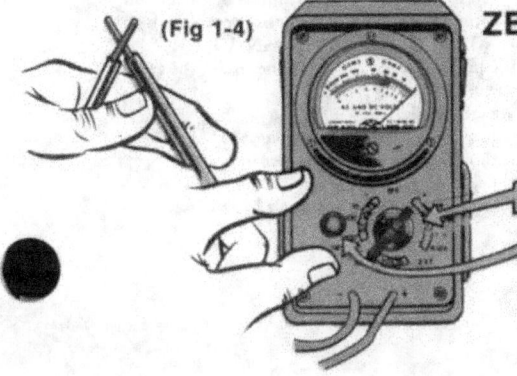

(Fig 1-4)

ZEROING AN/URM-105

Needle should be over "0" after Steps 1 and 2.
1. Put selector switch on X1 OHMS.
2. Touch the probes together while turning OHMS ADJ knob until needle is over "0" on right side of scale.

SELECTOR SWITCH

OHMS ADJ KNOB

If the needle will not zero, replace the batteries. If the needle still will not zero with new batteries, turn the meter in for repair.

1

ZEROING TS-352B/U

1. Put RANGE switch on RX1.
2. Put small black probe in this jack.
3. Put small red probe in this jack.
4. Put FUNCTION switch on OHMS.
5. Touch long probes together . . .

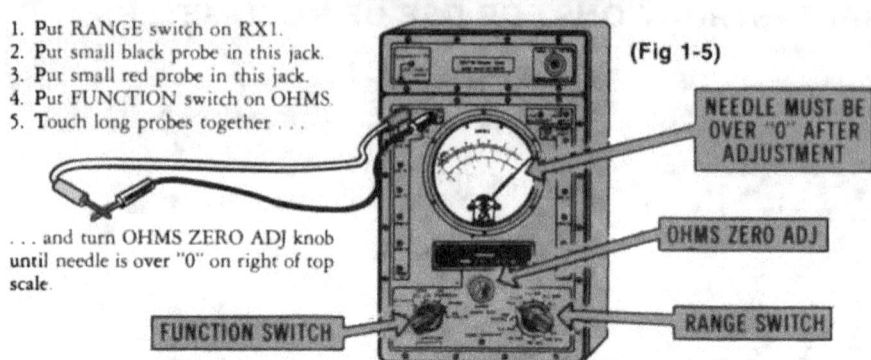

(Fig 1-5)

NEEDLE MUST BE OVER "0" AFTER ADJUSTMENT

OHMS ZERO ADJ

. . . and turn OHMS ZERO ADJ knob until needle is over "0" on right of top scale.

FUNCTION SWITCH

RANGE SWITCH

If needle will not "zero," replace the batteries. If the needle still will not "zero" after replacing the batteries, turn the meter in for repair.

ZEROING SIMPSON 160

1. Put Selector switch on RX1.
2. Put black probe in COM jack.
3. Put red probe in + jack.
4. Put the polarity switch in the up position.
5. Touch the probes together and turn the OHMS ADJ knob until the needle is over the "0" on the right side of the top scale.

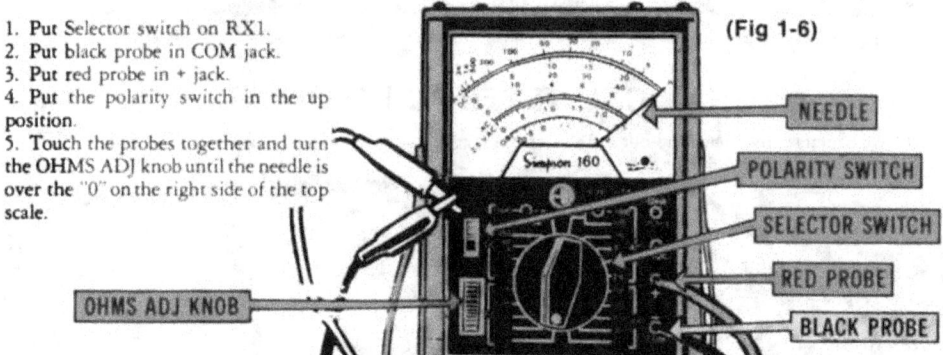

(Fig 1-6)

NEEDLE

POLARITY SWITCH

SELECTOR SWITCH

RED PROBE

BLACK PROBE

OHMS ADJ KNOB

If needle will not "zero, replace the batteries. If the needle still will not "zero" after replacing the batteries, turn the meter in for repair.

TESTING FOR CONTINUITY

Continuity tests are made to check for breaks in a circuit (such as the electrical cable illustrated). To make a continuity check, do the following steps:

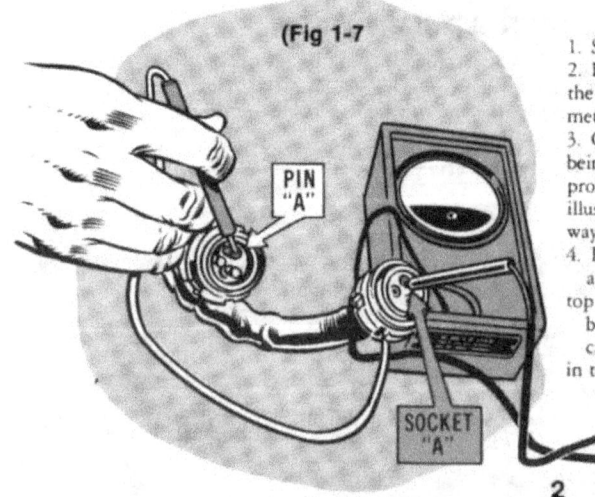

(Fig 1-7)

PIN "A"

SOCKET "A"

1. Set up for ohms (RX1, X1) and "zero" the meter.
2. Disconnect the circuit being tested. To be safe, disconnect the battery ground strap. Failure to do this can damage the meter.
3. Connect the meter probes to both terminals of the circuit being tested. In continuity testing, it makes no difference which probe goes on which terminal. (The AN/URM-105 is illustrated, but the probes are connected to the circuit the same way with all three multimeters.)
4. Look at the meter needle.
   a. If the needle swings to the far right over the "0" on the top scale (on all three multimeters) the circuit is good.
   b. If the needle doesn't move, the circuit is bad.
   c. If the needle jumps or flickers, there is a loose connection in the circuit.

2

TESTING FOR SHORTS

A short (or short circuit) occurs when two circuits that should not be connected have metal to metal contact with each other. A short also occurs when a circuit that should not touch ground has metal-to-metal contact with ground. Remember, the entire vehicle chassis is part of a circuit—the "Ground" circuit. To check for shorts, do the following steps:

1. Set up for ohms (RX1, X1) and "zero" the multimeter.
2. Disconnect the circuit being tested. To be safe, disconnect the battery ground strap (failure to do this can damage the meter).

3. With any of the three multimeters, connect one probe to one circuit and the other probe to the other circuit. The example below shows a check to see if wire "A" is shorted to wire "B" in the wiring harness.

SOCKET "B"

SOCKET "A"

(Fig 1-9)

4. Look at the needle.
  a. If the needle swings to the far right over the "O" on the top scale the circuits are shorted.

b. If the needle doesn't move, the circuits are good.
c. If needle jumps or flickers, the circuits are occasionally shorted.

TESTING RESISTANCE

To measure resistance in a circuit, do the following steps:
1. Set up for ohms (RX1, X1) and "zero" the multimeter.
2. Disconnect the circuit being tested. To protect the meter disconnect the battery ground strap.

3. With any of the three multimeters, connect the probes to the circuits or items to be measured.

AN/URM-105

(Fig 1-10)

TS-352 B/U

NEEDLE READS "38 OHMS" ON TOP SCALE OF EACH

4. Read the meter. With the meter switch on "RX1" or "X1" range, the reading is taken from the top scale, reading from right to left.

USING THE DC VOLTS SCALE

The DC Volts Scale is used to measure all voltages on the vehicle.
Before using the multimeter to measure DC voltage, do the following steps that match the multimeter you have:

AN/URM-105

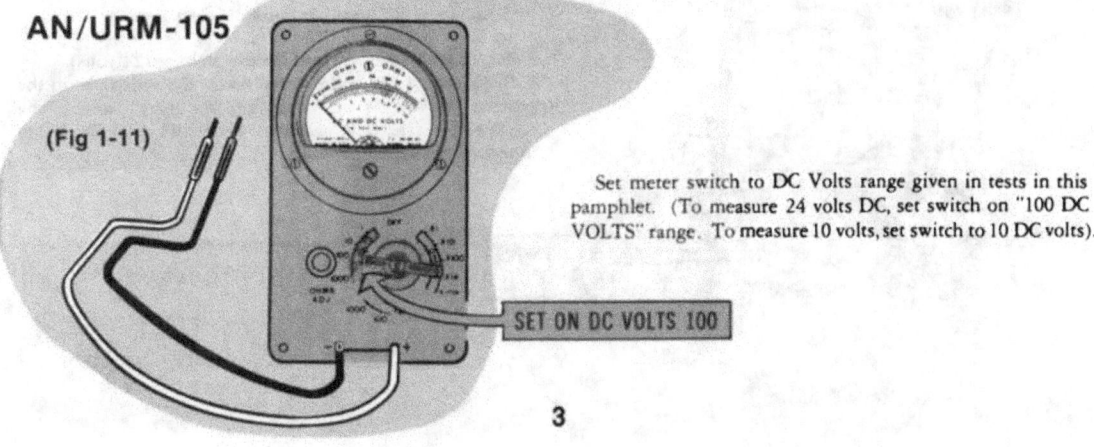

(Fig 1-11)

SET ON DC VOLTS 100

Set meter switch to DC Volts range given in tests in this pamphlet. (To measure 24 volts DC, set switch on "100 DC VOLTS" range. To measure 10 volts, set switch to 10 DC volts).

3

TS-352 B/U

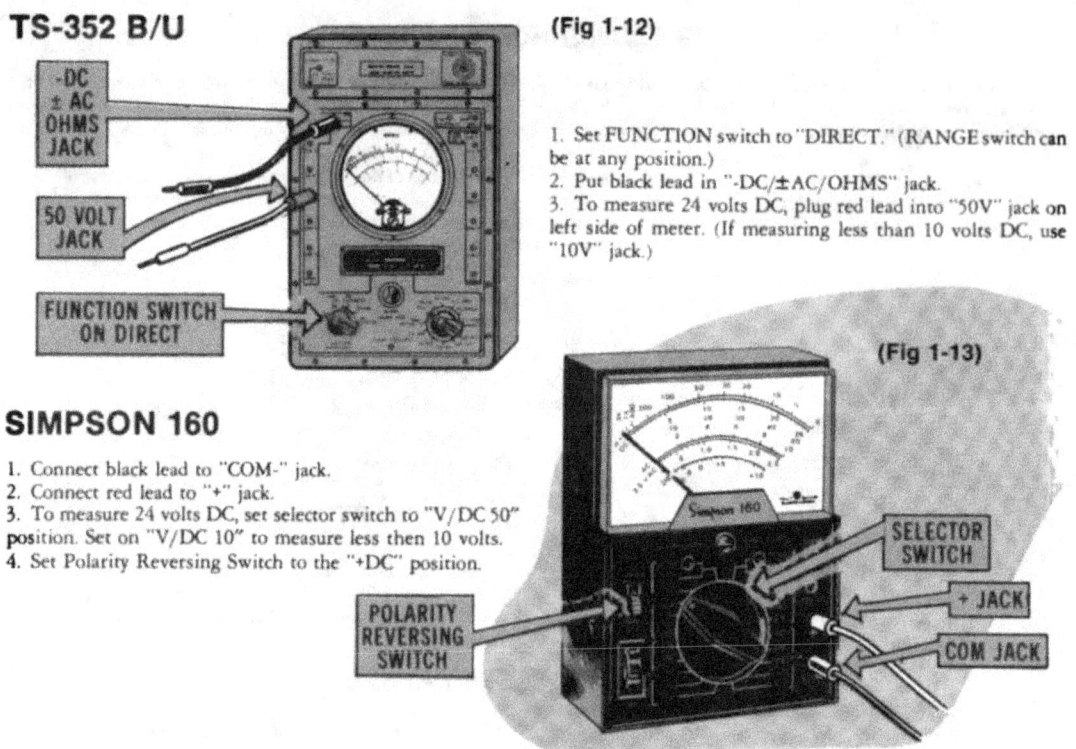

(Fig 1-12)

1. Set FUNCTION switch to "DIRECT." (RANGE switch can be at any position.)
2. Put black lead in "-DC/±AC/OHMS" jack.
3. To measure 24 volts DC, plug red lead into "50V" jack on left side of meter. (If measuring less than 10 volts DC, use "10V" jack.)

SIMPSON 160

1. Connect black lead to "COM-" jack.
2. Connect red lead to "+" jack.
3. To measure 24 volts DC, set selector switch to "V/DC 50" position. Set on "V/DC 10" to measure less then 10 volts.
4. Set Polarity Reversing Switch to the "+DC" position.

(Fig 1-13)

MEASURING DC VOLTAGE

(BATTERY VOLTS)

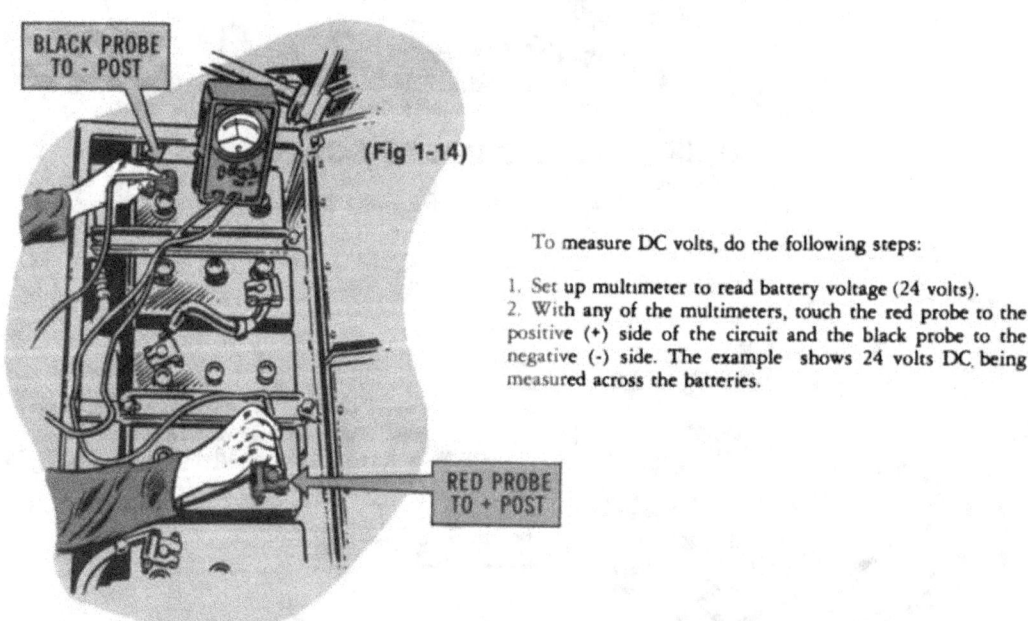

(Fig 1-14)

To measure DC volts, do the following steps:

1. Set up multimeter to read battery voltage (24 volts).
2. With any of the multimeters, touch the red probe to the positive (+) side of the circuit and the black probe to the negative (-) side. The example shows 24 volts DC being measured across the batteries.

4

Here're examples of other DC voltage tests—reading each of the scales:

TS-352 B/U

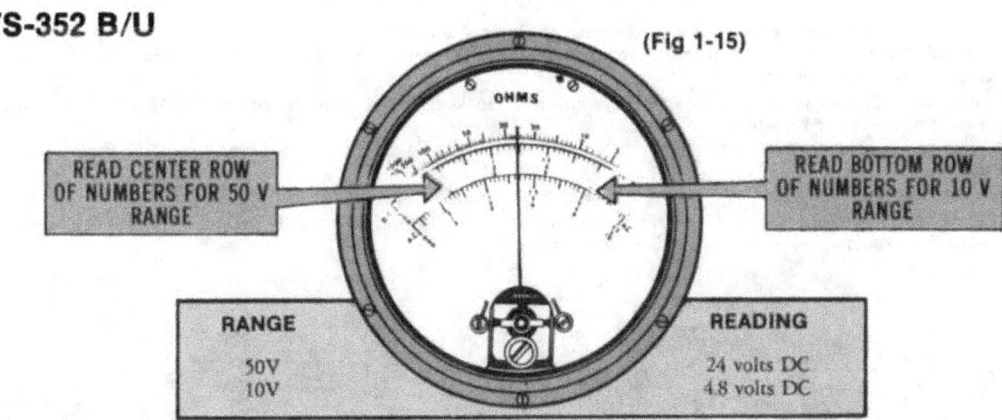

(Fig 1-15)

READ CENTER ROW OF NUMBERS FOR 50 V RANGE

READ BOTTOM ROW OF NUMBERS FOR 10 V RANGE

RANGE	READING
50V	24 volts DC
10V	4.8 volts DC

AN/URM-105

Read the upper, black straight-lined portion of the "AC and DC volts" scale.

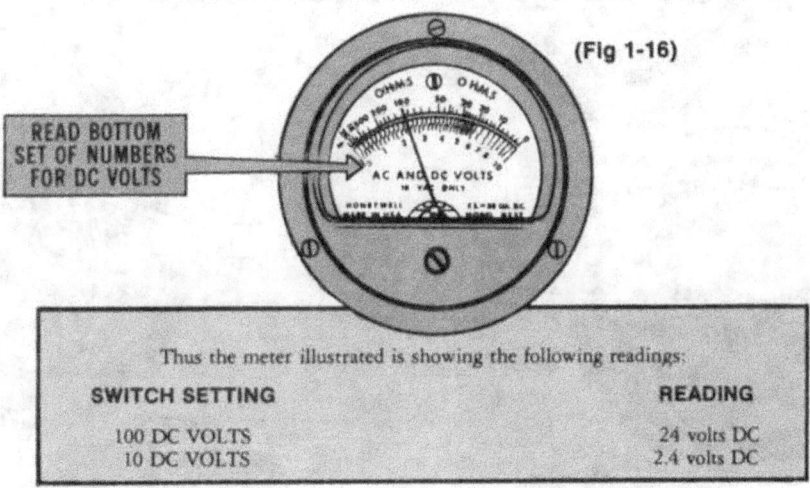

(Fig 1-16)

READ BOTTOM SET OF NUMBERS FOR DC VOLTS

Thus the meter illustrated is showing the following readings:

SWITCH SETTING	READING
100 DC VOLTS	24 volts DC
10 DC VOLTS	2.4 volts DC

SIMPSON 160

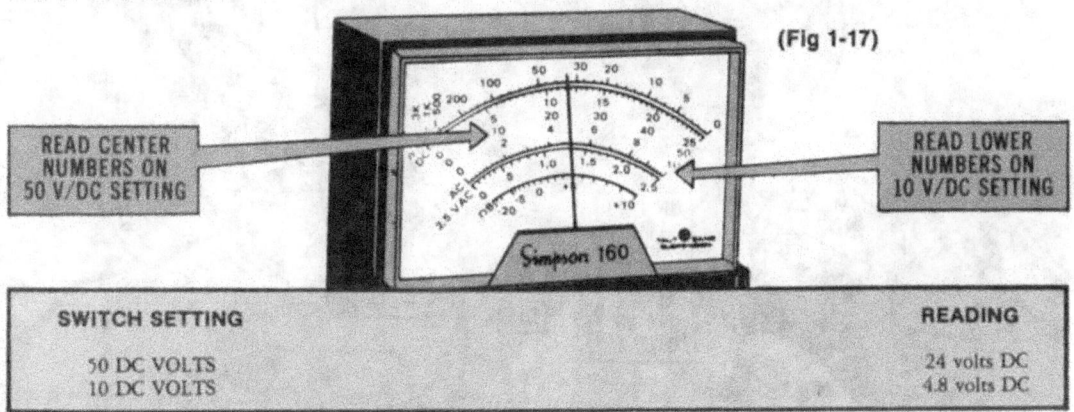

(Fig 1-17)

READ CENTER NUMBERS ON 50 V/DC SETTING

READ LOWER NUMBERS ON 10 V/DC SETTING

SWITCH SETTING	READING
50 DC VOLTS	24 volts DC
10 DC VOLTS	4.8 volts DC

PIN AND SOCKET IDENTIFICATION

Knowing what pins or sockets to look for on various connectors can be sticky.

To simplify the search, look for the keyway in the connector attached to the cable and look for the key in the connector attached to the component (regulator, etc.).

If the pins or sockets are offset from the key or keyway, the "A" socket is the one on the left of the key or keyway. The "A" pin is on the right of the key or keyway.

To identify other pins or sockets, you must eyeball the connector and find the letter of the pin or socket. The letter is stamped next to each.

Following are locations of pins and sockets in typical connectors which you will test in this pamphlet:

NOTE: *Bad pins or sockets are possible causes of charging system problems. If the pins or sockets are* burned, bent *or* pushed back *into the connector, repair or replace them as necessary.*

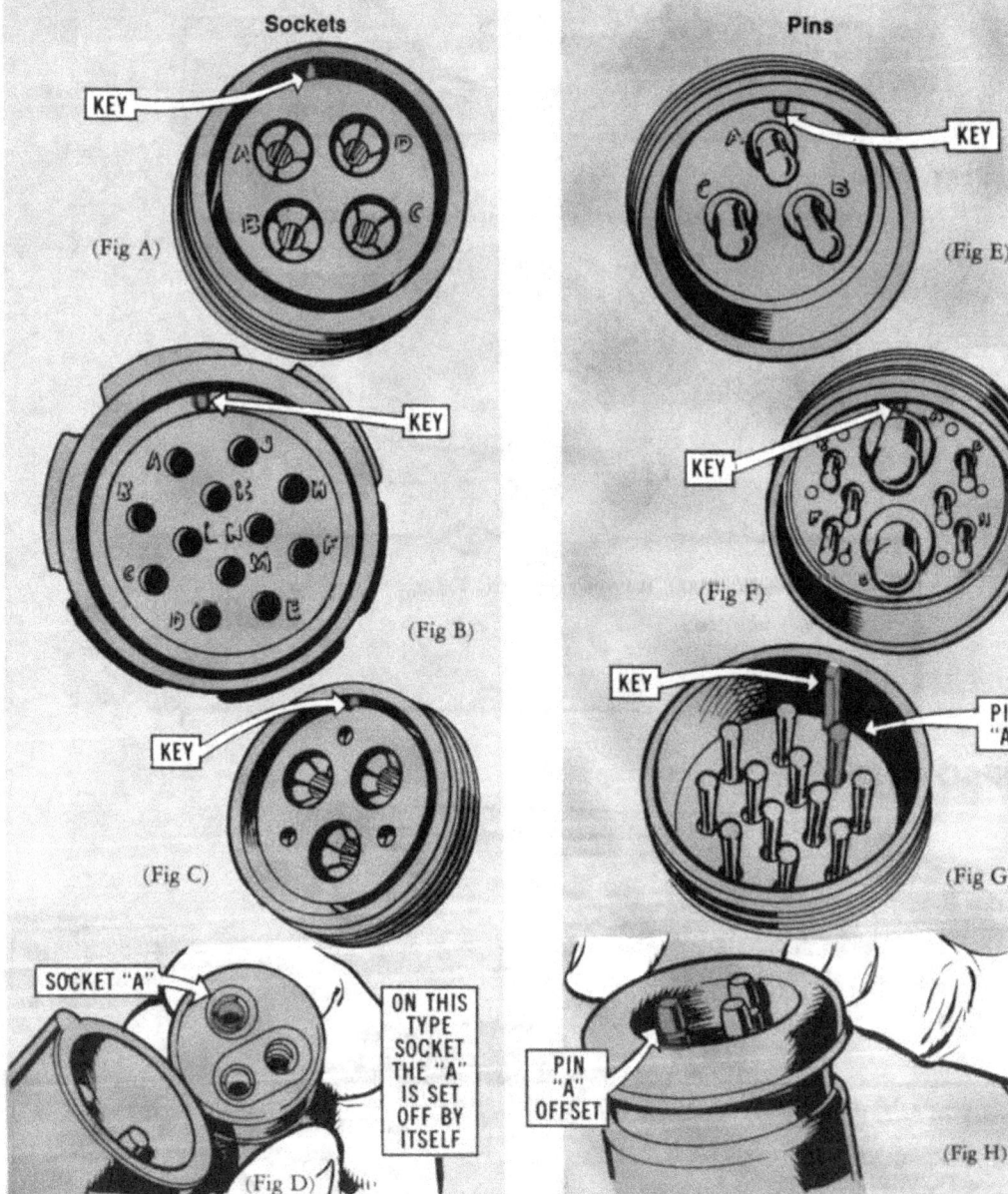

USING THE OPTICAL ANTIFREEZE/BATTERY TESTER

Test the specific gravity of electrolyte in each battery cell with Optical Antifreeze/Battery Tester NSN 6630-00-105-1418. Instructions for its use are printed below. To be sure the tester is working properly, put several drops of clean water (tap or whatever) on the measuring window with the dipstick. The tester's OK if you get a +30 to +34 F reading on the right hand scale. You do not have to adjust for temperature when taking any reading.

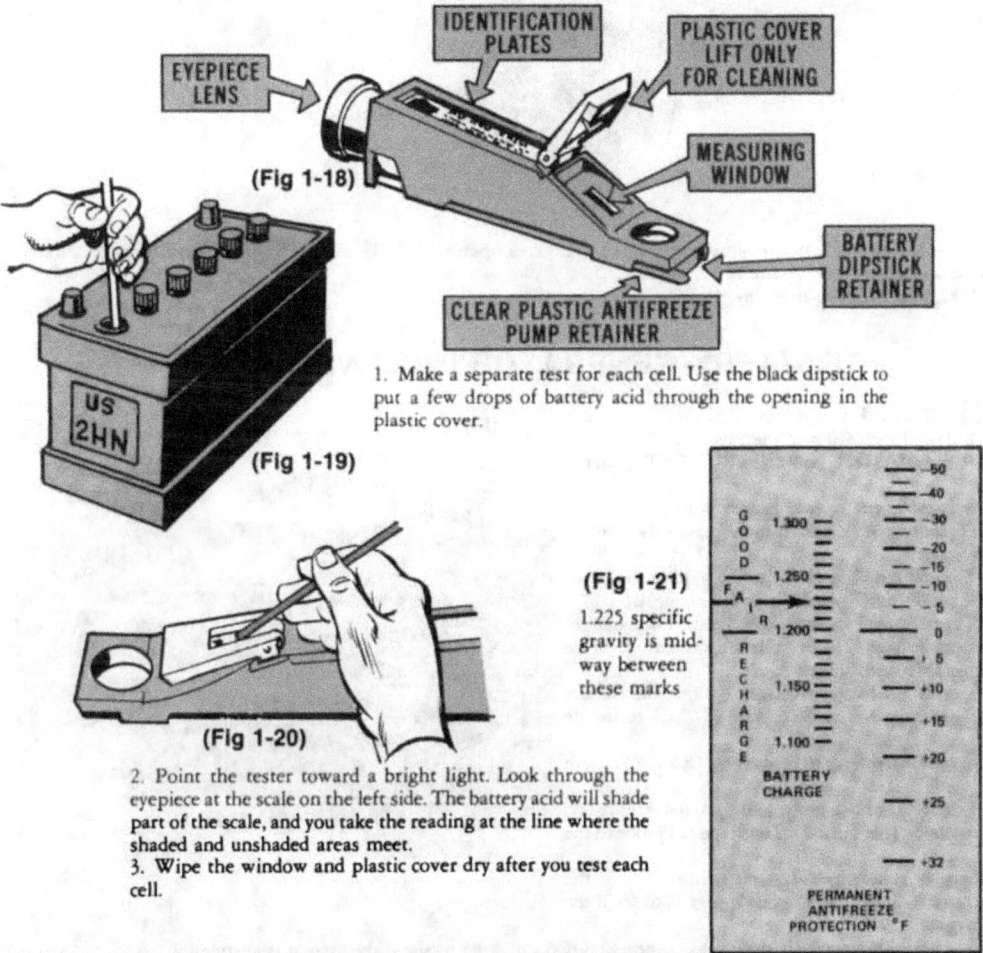

EYEPIECE LENS

IDENTIFICATION PLATES

PLASTIC COVER LIFT ONLY FOR CLEANING

(Fig 1-18)

MEASURING WINDOW

BATTERY DIPSTICK RETAINER

CLEAR PLASTIC ANTIFREEZE PUMP RETAINER

US 2HN

(Fig 1-19)

1. Make a separate test for each cell. Use the black dipstick to put a few drops of battery acid through the opening in the plastic cover.

(Fig 1-20)

(Fig 1-21)

1.225 specific gravity is midway between these marks

2. Point the tester toward a bright light. Look through the eyepiece at the scale on the left side. The battery acid will shade part of the scale, and you take the reading at the line where the shaded and unshaded areas meet.
3. Wipe the window and plastic cover dry after you test each cell.

BATTERY CHARGE

PERMANENT ANTIFREEZE PROTECTION °F

(Caution: Keep the plastic cover snug against the window when you insert the battery acid and while you're taking the reading. If the dividing line of the shaded area isn't sharp, re-clean the window and plastic cover, seat the cover carefully, and repeat the test.)

TOOLS LIST

Following is a list of tools you'll need to perform charging system tests on your vehicle:
(Note: Depending on your vehicle, you may need all or most of those listed.)

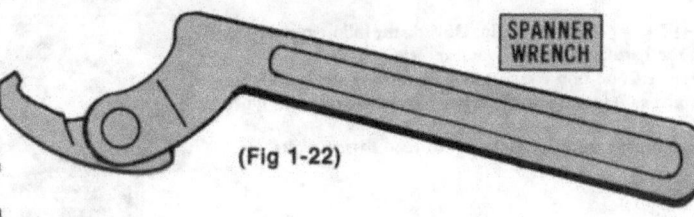

SPANNER WRENCH

°Flashlight
°Allen Wrench, ¼-in
°Spanner Wrench, ¼-2-in
°Open End Wrenches
  ¼-in, ⅜-in, ⁷/₁₆-in, ½-in, ⁹/₁₆-in, ¼-in
°Screwdrivers, Flat Tip, 4-in, 8-in
°Screwdriver, Phillips, Cross Tip , 6-in

(Fig 1-22)

7

Chapter 2

CHARGING CIRCUIT MALFUNCTIONS

WARNING: When testing electrical circuits or components, do not wear rings or watches.

(Fig 2-1)

BATTERY
GENERATOR
INDICATOR

1. If the battery-generator indicator needle does not move in any area when MASTER and INSTRUMENT switches are ON, and all other lights and gages work, do Check #1, page 8.
2. Do Check #2, page 8, if you have any charging system problems.

BATTERY-GENERATOR INDICATOR

CHECK #1-This check determines if the battery-generator indicator is bad or if the wiring to the gage is bad.

Step A. Turn MASTER and INSTRUMENT switches OFF.

Step B. Set multimeter to read battery volts.

Step C. Disconnect Wire #27 from the rear of the battery-generator indicator.

Step D. Ground the black probe to the vehicle and touch the red probe to Wire #27 which you just disconnected.

Step E. Turn "power" switches ON. Observe the meter.

1. If battery voltage (23-26) volts are present, turn the power switches OFF and do Step F.

2. If there is no battery voltage, remove the battery ground strap and repair or replace Wire #27 back to the tie point.

Step F. Set the multimeter to read ohms, (RX1, X1). Zero the meter.

Step G. Connect the black probe to the ground terminal on the battery-indicator (See Fig 2-2). Touch the red probe to the vehicle hull.

1. If there is continuity (zero resistance) replace the battery-generator indicator. Do Check #5, page 9, to see if the system is charging.

2. If there is resistance, repair the ground connection. Do Check #5 to see if the system is charging.

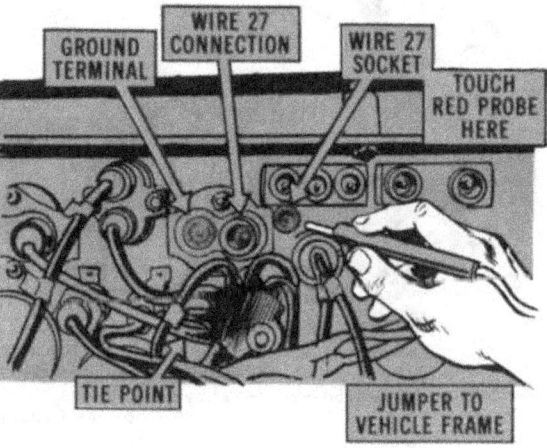

(Fig 2-2)

GROUND TERMINAL

WIRE 27 CONNECTION

WIRE 27 SOCKET

TOUCH RED PROBE HERE

TIE POINT

JUMPER TO VEHICLE FRAME

——————————— End Check #1 ———————————

BATTERY VOLTAGE

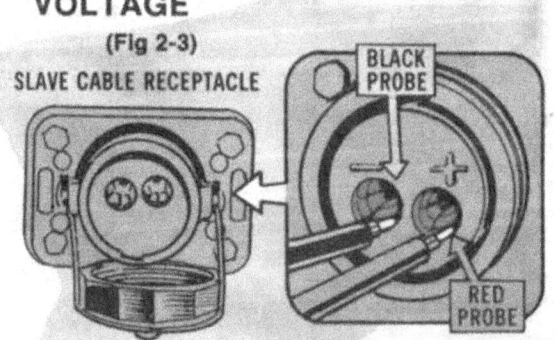

(Fig 2-3)

SLAVE CABLE RECEPTACLE

BLACK PROBE

RED PROBE

CHECK #2-With the engine OFF, do the following steps to check the battery voltage of your vehicle:

Step A. Condition the batteries by turning the headlights ON high beam for 30 seconds. After 30 seconds, turn the lights OFF.

Step B. Set the multimeter up to read battery volts.

8

Step C. Touch the black probe to the negative (-) terminal of the battery or to the negative (-) terminal of the slave cable receptacle. Touch the red probe to the positive (+) terminal of the battery or to the positive (+) terminal of the slave cable receptacle.

    1. If the meter reading is 23-26 volts, do Check #3.

    2. If the meter reading is less than 23 volts, test the batteries by doing Check #4.

———————————————————— End Check #2 ————————————————————

BATTERY CRANKING VOLTAGE

CHECK #3-To test for battery cranking voltage drop, do the following steps:

(NOTE: For Check #3 on spark plug system, leave the ignition switch OFF. On diesel system, pull the engine STOP knob to the stop position.)

Step A. Set multimeter to read battery volts.

Step B. Touch the black and red probes to the battery or slave receptacle terminals as you did in Step C of Check #2 and watch the needle.

Step C. Have someone crank the engine for up to 10 seconds (no longer).

    1. If the vehicle does not crank, test the batteries (Check #4). If the batteries check out OK, troubleshoot your starter circuit. When this has been corrected, do Check #5.

    2. If the reading drops below 18 volts during cranking, do Check #4.

    3. If the meter reading stays at or above 18 volts during cranking, do Check #5.

———————————————————— End Check #3 ————————————————————

BATTERY SPECIFIC GRAVITY

CHECK #4-The following steps will let you know the condition of your batteries:

See page   , for use of the battery/antifreeze tester.

WARNING: If you find that you need to clean and tighten the battery posts, disconnect the ground strap first. Hook the ground strap back up when you finish. Also remember when removing batteries, disconnect the ground strap first. When installing batteries, connect the ground strap last. When working with batteries, it is always a good idea to use safety goggles.

Step A. Give the batteries a good inspection. Look for loose or corroded connections on the batteries and at the ground strap vehicle connection. You can tell if the connections are tight by trying to move the connection by hand. Do not apply excessive pressure on the cables.

    1. If the connections are clean and tight, do Step B.

    2. If the connections are corroded or loose, clean them and apply a light coat of grease or corrosion inhibitor. Retighten the connections and do Check #3.

Step B. Use the battery/antifreeze tester to measure the strength of your battery. Test every cell. Writing down the readings will help you keep track. You may want to make a sketch of the batteries and write the readings down on the sketch.

    1. If any cells read less than 1.225 or if the readings vary more than .025 (2½ divisions), do Step C. (See Fig 1-21.)

    2. If all the cells read above 1.225 and do not vary more than .025 (2½ divisions), the battery checks OK. Do Check #5.

Step C. Attempt to recharge the batteries. The sooner you get a discharged battery recharged, the better. Batteries left discharged for long periods of time are very hard to recharge.

    1. If the battery comes back up to charge and the cells do not vary more than .025 (2½ divisions), do Check #5.

    2. If the battery will not come back up or the cells still vary by more than .025, replace the battery. Do Check #5.

(NOTE: TM 9-6140-200-12, dtd Sep 73, will clue you in on how to pull maintenance on your batteries.)

———————————————————— End Check #4 ————————————————————

BATTERY VOLTAGE RISE

CHECK #5-To test for battery charging voltage rise, follow these steps:

Step A. Start the vehicle, turn on high beam lights and allow engine to reach operating temperature.

Step B. Set the multimeter up to read battery volts.

Step C. Touch the black and red probes to the battery or slave receptacle terminals as you did in Step C, Check #2. Watch needle.

Step D. Have someone increase the engine speed to 1000-1200 RPM and hold steady.

    1. If the needle reading is steady in the 27-29 volts range and there are no complaints about the vehicle, the charging system is OK.

    2. If the needle reading is steady in the 27-29 volt range, but there are complaints (batteries go weak, use excessive water, get hot) troubleshoot the vehicle charging system. See Table of Contents for your particular charging system tests.

    3. If the needle reading is above or below the 27-29 volt range, troubleshoot the vehicle charging system. See Table of Contents for your particular charging system tests.

———————————————————— End Check #5 ————————————————————

9

Chapter 3

25 AMP CHARGING SYSTEM
(WHEEL VEHICLE)

Note: Do the Battery Voltage Checks on pages 8 and 9 before doing Test 1.

Test 1- Do the following steps to test the reverse current relay inside the voltage regulator (a bad reverse current relay will cause your batteries to run down over night):

Step A. With the engine off, disconnect the regulator-to-generator cable at the generator gooseneck.

(Fig 3-1)

GENERATOR GOOSENECK

Step B. Set the multimeter up to read battery volts.

Step C. Touch the black probe to a good ground and touch the red probe to pin A of the disconnected cable.

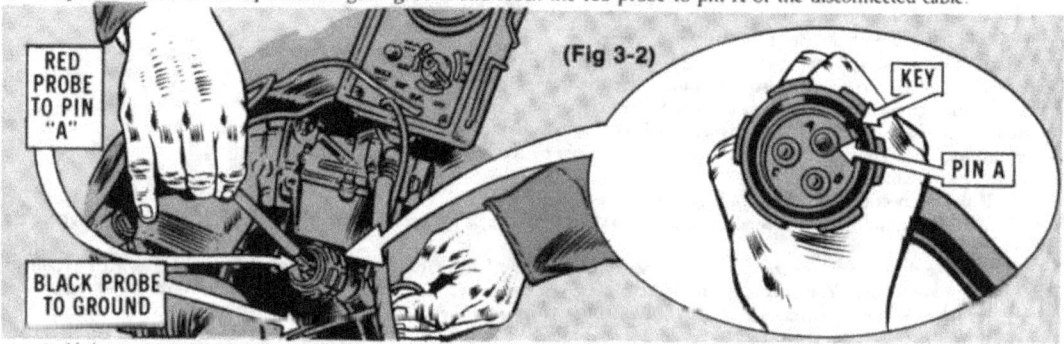

RED PROBE TO PIN "A"

BLACK PROBE TO GROUND

(Fig 3-2)

KEY

PIN A

1. If the meter reads 0 volts, the reverse current relay is OK. Go on to Test 2.

2. If the meter reads any voltage, the regulator reverse current relay is bad. Replace the regulator. Repeat Check 5 on page 9 to be sure that the new regulator you installed works OK.

Test 2- Do the following steps to test the generator armature and field:

Step A. Set the multimeter up to read ohms, (RX1, X1). "Zero" the meter.

Step B. Touch the black probe to a good ground and touch the red probe to socket A in the gooseneck.

(Fig 3-3)

SOCKET "A"

1. If the meter needle swings to the right and reads 0-4 ohms on the ohms scale, do Step C.

2. If the meter reads more than 4 ohms, replace the generator and repeat Check 5 on page 9 to make sure the system is charging right.

Step C. Touch the black probe to a good ground and touch the red probe to socket B in the gooseneck connector.

(Fig 3-4)

SOCKET "B"

1. If the meter needle swings to 20-45 ohms on the ohms scale, do Test 3.
2. If the needle is not in the 20-45 ohms range, replace the generator and repeat Check 5. page 9.

Test 3- Do the following steps to test the generator for residual voltage:

Step A. Remove the short red probe of the multimeter from the OHMS jack and place it in the 10 volt jack. Set the meter FUNCTION switch to DIRECT.
Step B. Start the engine and let it idle between 1,000-1,200 RPM.
Step C. Touch the black probe to a good ground. Touch the red probe to socket A in the gooseneck connector.

(Fig 3-5)

KEY

SOCKET "A"

1. If meter reads between .5 and 4 volts, the generator is okay. Turn the engine off. Do Test 4.
2. If the reading is more than 4 volts, turn the engine off and replace the generator. Repeat Check 5 on page
3. If the reading is less than .5 volts or the needle deflects to the left of 0, do Step D to polarize the generator.

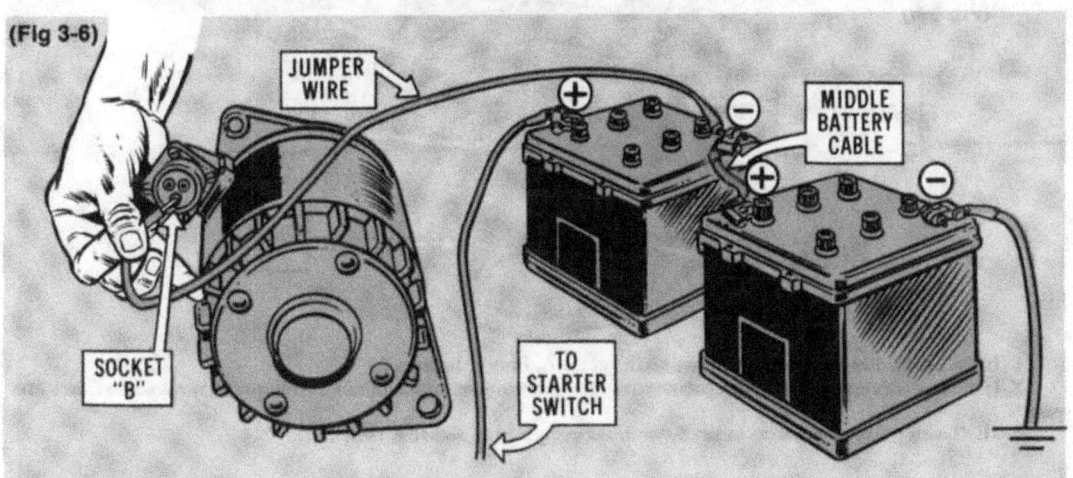

(Fig 3-6)

JUMPER WIRE

MIDDLE BATTERY CABLE

SOCKET "B"

TO STARTER SWITCH

Step D. Polarizing the generator: With the engine and ignition switch off, use an insulated jumper wire to jump from the middle battery cable (either terminal) to socket B in generator gooseneck. Repeat Steps B and C. If the reading is still bad (less than .5 volt or deflects left), repeat Step D. Do not attempt to polarize the generator more than twice. If the reading is bad after the second attempt replace the generator.

11

Test 4 Do the following steps to test the generator-to-regulator cable for continuity and shorts:
Step A. Disconnect the generator-to-regulator cable at the regulator connector (the generator end is already disconnected).

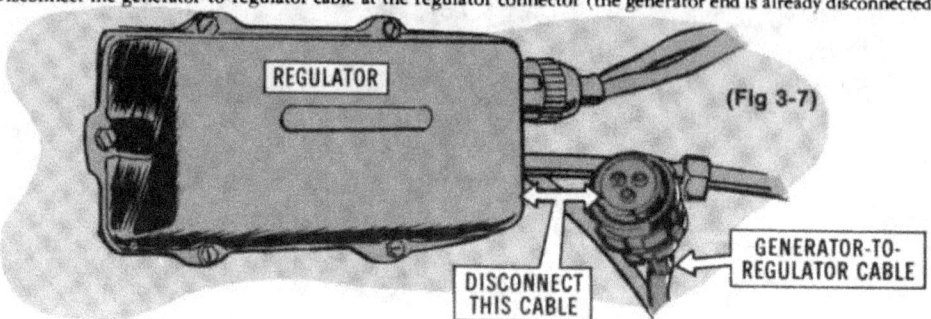

REGULATOR

(Fig 3-7)

DISCONNECT
THIS CABLE

GENERATOR-TO-
REGULATOR CABLE

NOTE: *Check the pins and sockets on both ends of the cable. If they are burnt, bent or pushed back, repair or replace them.*

Step B. Set the meter to read ohms (RX1, X1). "Zero" the meter.
Step C. Touch the red and black probes to the matching connections on each end of the cable (pin A to socket A, pin B to socketB). The C pins do not have to be checked for this test.

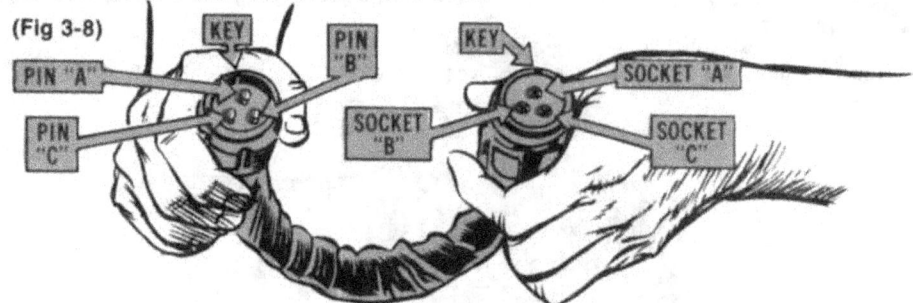

(Fig 3-8)

KEY   PIN "B"   KEY   SOCKET "A"

PIN "A"   PIN "C"   SOCKET "B"   SOCKET "C"

1. If the meter reads 0 ohms each time, do Step D.
2. If the meter does not read 0 ohms each time (the cable is broken), repair or replace the generator-to-regulator cable.
Step D. Touch the black probe to the outside of the cable connector.   Touch the red probe to sockets A, B and C.
Note the meter reading each time.

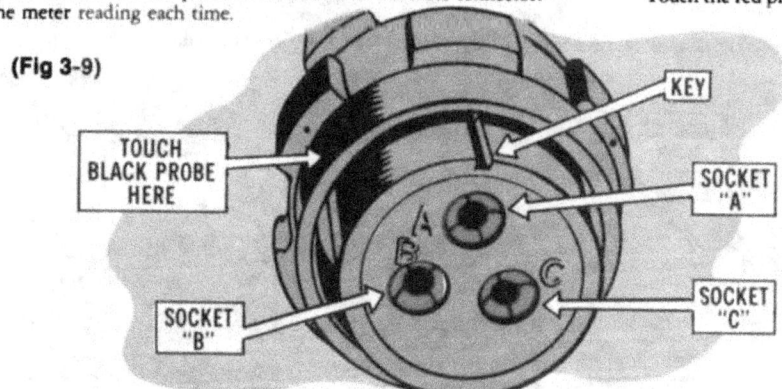

(Fig 3-9)

KEY

TOUCH
BLACK PROBE
HERE

SOCKET "A"

SOCKET "B"

SOCKET "C"

1. If the needle does not move when you touch sockets A, B or C do Step E.
2. If the needle swings to the right, repair or replace the generator-to-regulator cable. After you fix the cable, do Check 5 on page 9.
Step E. Touch the black probe to socket B, touch the red probe to socket A then C.

1. If the needle does not move, do Test 5.
2. If the needle moves (the cable is shorted), repair or replace the cable. Do Check 5 on page 9.

Test 5 Do the following steps to test the regulator-to-battery cable:

Step A. Disconnect the regulator-to-battery cable at the regulator. Caution: Pin A or C is hot. Don't touch to vehicle chassis.

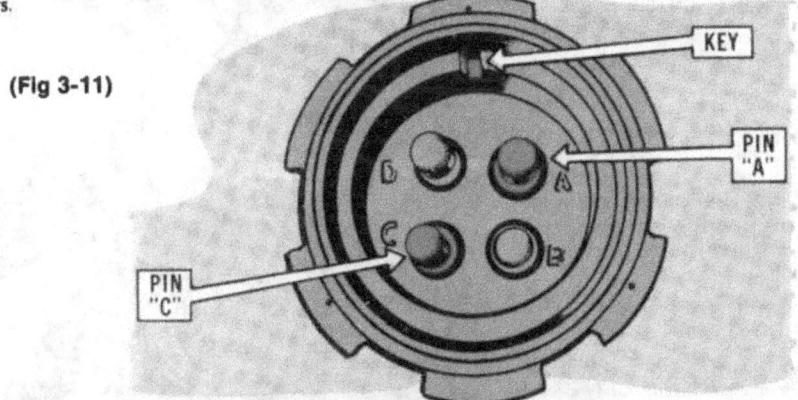

(Fig 3-10)

CONNECTION

REGULATOR

REGULATOR-TO-BATTERY CABLE

Step B. Set up the meter for battery volts.

Step C. Touch the black probe to a good ground. In the cable connector touch the red probe to pin C on ¼ Ton, pin A on all others.

(Fig 3-11)

KEY

PIN "A"

PIN "C"

1. If the meter reading is 23-26 volts, replace the regulator. Repeat Check 5 on page 9.
2. If the meter reading is less than 23 volts, do Step D.

Step D. Clean the battery posts, terminals, and the battery ground connection. Repeat Step C.

1. If the voltage now reads 23-26 volts; repeat Check 5 on page 9.
2. If the meter reads less than 23 volts, get Direct Support to repair or replace the regulator-to-battery cable. Repeat Check 5 when the cable is repaired or replaced.

END OF 25 AMP CHARGING SYSTEM TESTS
(WHEEL VEHICLES)

Chapter 4

60 AMP CHARGING SYSTEM

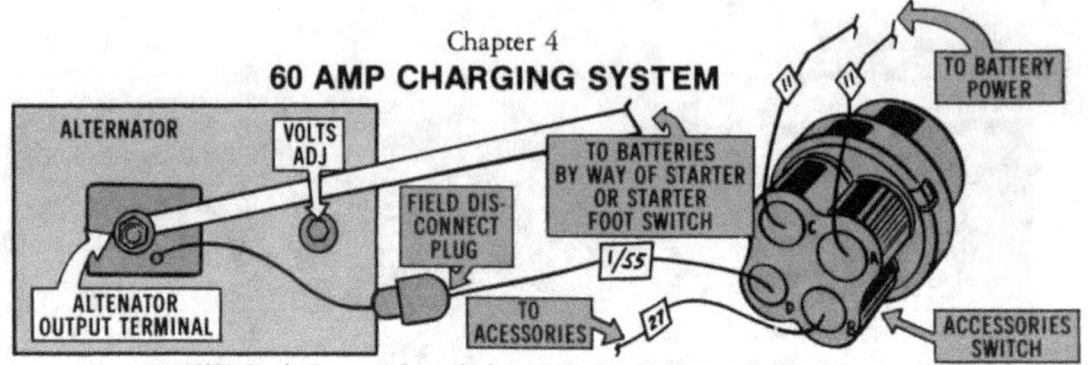

NOTE: *Do the Battery Voltage Checks on pages 8 and 9 before doing Test 1.*

Test 1 To test for a complete circuit between the batteries and the alternator field and ignition switch, do the following steps:

Step A. Disconnect the alternator field plug.

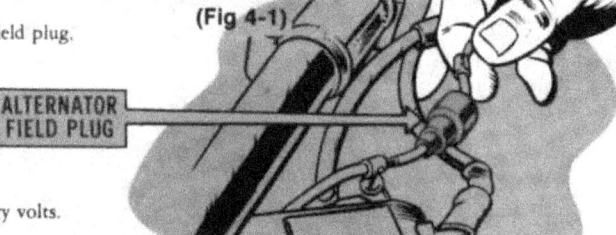

(Fig 4-1)

ALTERNATOR
FIELD PLUG

Step B. Set the meter to read battery volts.

Step C. With the ignition switch OFF, touch the black probe to a good ground and touch the red probe to the pin in the field disconnect plug (the cable goes into the cable harness).

  1. If the meter reads 0 volts, do step D.
  2. If the meter does not read 0 volts, do Test 5.

Step D. Turn the ignition switch ON, but do not start the engine. Touch the black probe to a good ground and touch the red probe to the pin as in Step C.

  1. If the meter reads 23 volts or more, turn the ignition switch off. Do Test 2.
  2. If the meter reads less than 23 volts, turn the switch off. Do Test 5.

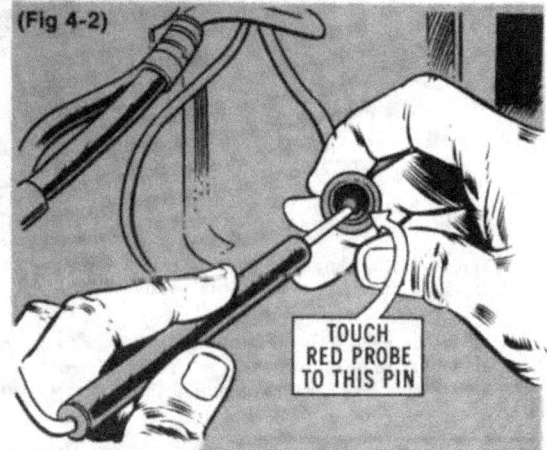

(Fig 4-2)

TOUCH
RED PROBE
TO THIS PIN

Test 2 To test the alternator field windings and the alternator rectifier do the following steps:

Step A. Set the meter to read ohms (RX1, X1). "Zero" the meter.

Step B. Touch the black probe to a good ground and touch the red probe to the socket in the field disconnect plug (the cable end that goes to the alternator).

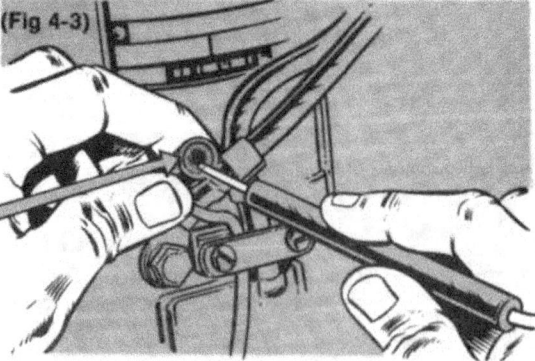

(Fig 4-3)

TOUCH RED PROBE
TO THIS SOCKET

  1. If the needle reads 200 to 500 ohms, do Step C.
  2. If the needle does not read 200 to 500 ohms, replace the alternator.

14

Step C. Disconnect the battery ground strap.

Step D. Remove the alternator terminal cover. Some alternators have rubberized insulation under the cover. Carefully remove this—be careful not to cut the insulation on the field wire.

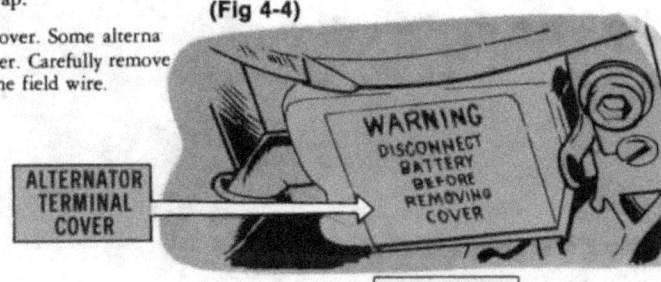
(Fig 4-4)

ALTERNATOR TERMINAL COVER

WARNING
DISCONNECT BATTERY BEFORE REMOVING COVER

Step E. Disconnect the alternator-to-battery cable at the alternator.

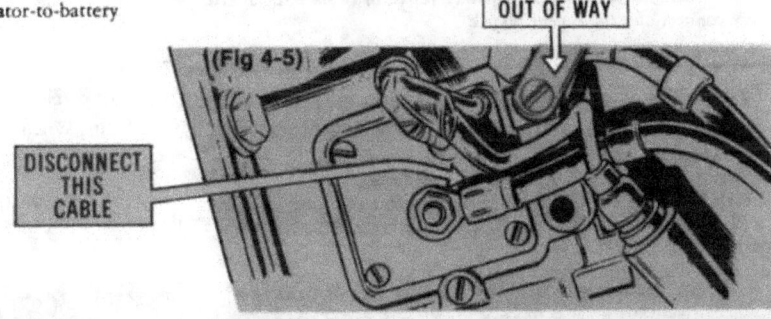
(Fig 4-5)

SWING CLAMP OUT OF WAY

DISCONNECT THIS CABLE

Step F. Touch the black probe to a good ground and touch the red probe to the alternator output terminal. Note the meter reading.

(Fig 4-6)

TOUCH RED PROBE TO THIS TERMINAL

Step G. Touch the red probe to a good ground and touch the black probe to the alternator output terminal. Note the meter reading.

1. If the meter reads 50 to 80 ohms in Step F and the needle does not move in Step G, the rectifier is okay. Do Test 3.

2. If the meter does not read 50 to 80 ohms in Step F or if the needle moves (a flicker or slight movement is OK) in Step G, the alternator rectifier is bad. Replace the alternator.

Test 3 Alternator-to-battery cable test:

Caution: Don't hook up batteries till after Step C.

Step A. Reconnect the alternator field plug.

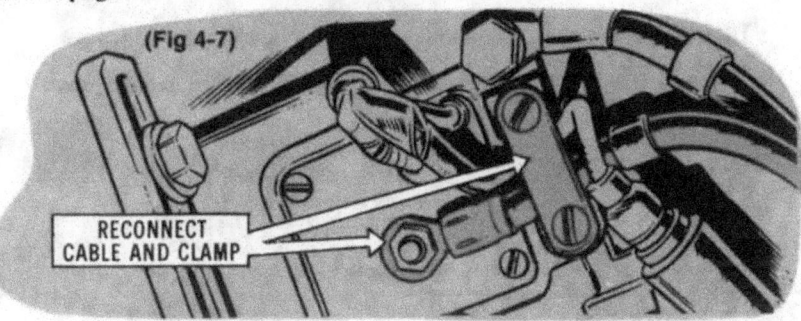
(Fig 4-7)

RECONNECT CABLE AND CLAMP

Step B. Reconnect the alternator-to-battery cable at the alternator.

Step C. Reconnect battery ground strap.
Step D. Set the meter to read battery volts.

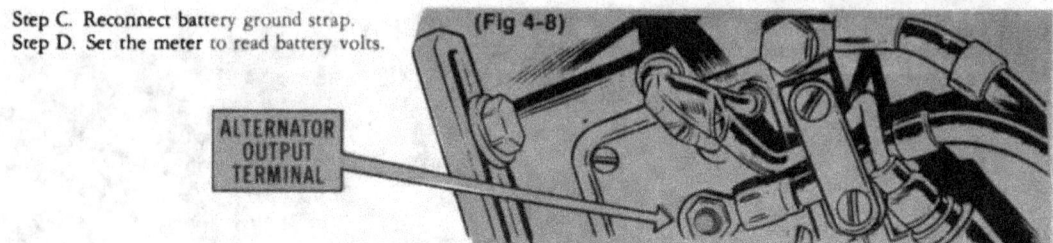
(Fig 4-8)

ALTERNATOR OUTPUT TERMINAL

Step E. Touch black probe to a good ground. Touch red probe to the alternator output terminal.

    1. If needle reads 23 to 26 volts, do Test 4.

    2. If needle reads less than 23 volts, repair or replace the alternator-to-battery cable. First, though, clean and tighten the battery connections and repeat Step E.

Test 4- Do the following steps to adjust the alternator output , if local SOP permits. If not replace the alternator.

Step A. Start engine. Idle at 1,0p0-1,200 RPM and turn on high beam lights. When engine reaches normal operating temperature, do Step B.

Step B. Set the meter up to read battery volts.

Step C. Touch black probe to a good ground and touch red probe to alternator output terminal.

    1. If the meter reads 27 to 29 volts the system checks OK. If the needle does not read 27 to 29 volts, adjust alternator output into the 27 to 28 volt range.

      (a) The voltage adjustment is at the back or front of the alternator. It is marked "VOLT ADJUST."

(Fig 4-9)

VOLT ADJUST

WARNING
DISCONNECT BATTERY BEFORE REMOVING COVER

      (b) Remove allen socket pipe plug at VOLT ADJUST. The plug covers the adjustment screw.

      (c) Turn the screw with a small flat tip screwdriver until you get a 27 to 29 volt reading.

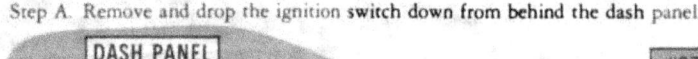
(Fig 4-10)

VOLTAGE ADJUST SCREW

    2. If you can't get a 27 to 29 volt reading, replace the alternator.

Test 5- Do the following steps to see if the ignition switch and its wiring are good:

Step A. Remove and drop the ignition switch down from behind the dash panel.

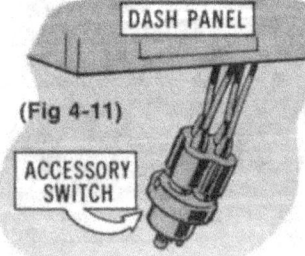

DASH PANEL

(Fig 4-11)

ACCESSORY SWITCH

NOTE: Wire letters A, B, C, and D are stamped on the end of the rubber connectors. If you cannot see the letters, use the locator key in Figure 4-12. Wire numbers are tagged on the leads.

Fig 4-12

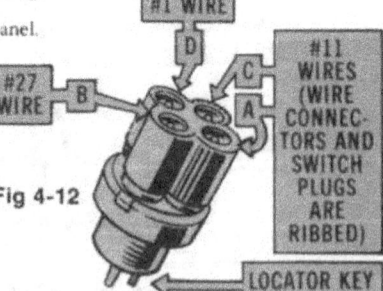

#1 WIRE
D
#27 WIRE
B
C
A
#11 WIRES (WIRE CONNECTORS AND SWITCH PLUGS ARE RIBBED)
LOCATOR KEY

16

Step B. Reconnect the switch handle. Disconnect Wire #1/55 from the switch.

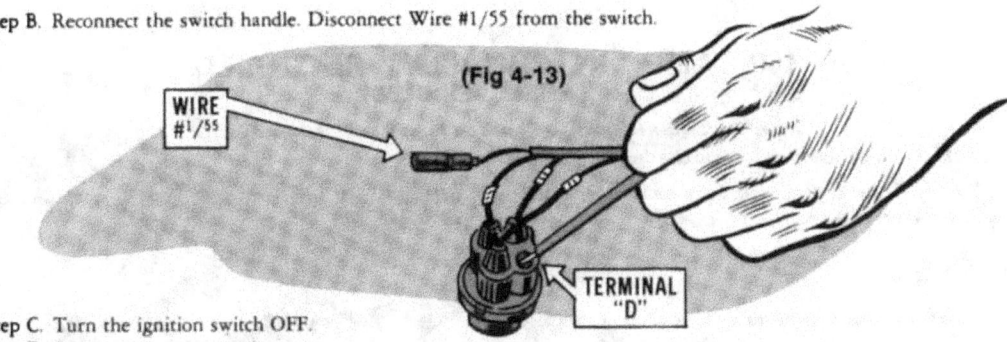

(Fig 4-13)

WIRE #1/55

TERMINAL "D"

Step C. Turn the ignition switch OFF.
Step D. Leave meter set up as is.
Step E. Touch black probe to a good ground and touch red probe to terminal D of the ignition switch (Terminal D is where you removed Wire #1/55).

(Fig 4-14)

1. If you get a voltage reading, replace the switch and repeat Check 5 on page 9.
2. If you get no voltage reading, do Step F.

Step F. Turn the ignition switch ON.
   1. If you get no voltage reading, turn the switch OFF and do Test 6.
   2. If the meter reads battery voltage (23-26 volts), Wire #1/55 is bad. Direct support must repair or replace the wiring harness.

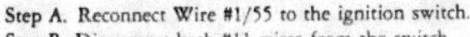

Test 6- Do the following steps to test the wiring between the ignition switch and the batteries:

Step A. Reconnect Wire #1/55 to the ignition switch.
Step B. Disconnect both #11 wires from the switch.

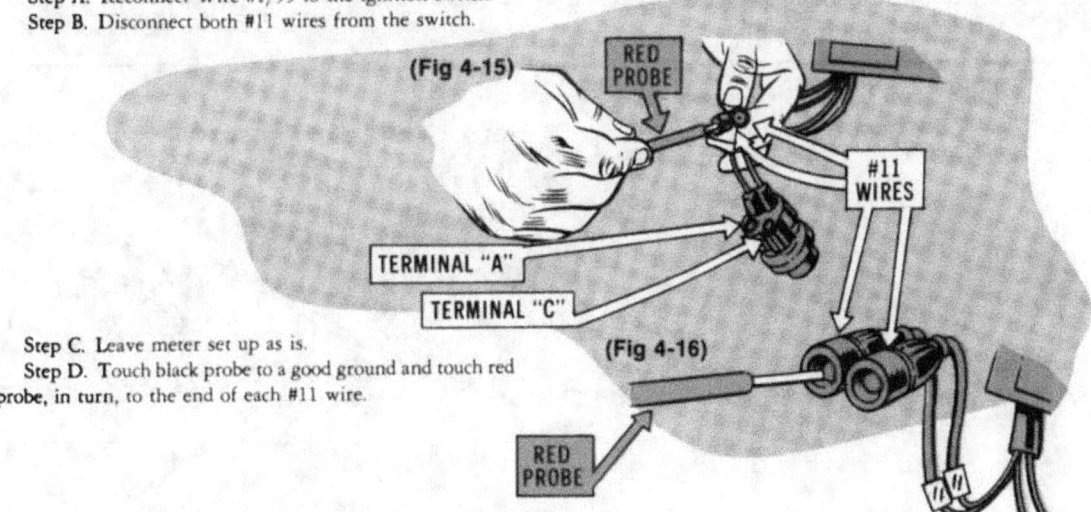

(Fig 4-15)

RED PROBE

#11 WIRES

TERMINAL "A"

TERMINAL "C"

(Fig 4-16)

RED PROBE

Step C. Leave meter set up as is.
Step D. Touch black probe to a good ground and touch red probe, in turn, to the end of each #11 wire.

1. If the needle reading is 23 to 26 volts at each #11 wire, replace the ignition switch and repeat Check 5, page
2. If you do not get a voltage reading at one or both #11 wires, have Direct Support repair or replace the wiring harness.

END OF 60 AMP CHARGING SYSTEM TESTS
(WHEEL VEHICLES)

Chapter 5
100 AMP INTERNALLY RECTIFIED
(WHEEL VEHICLES)

Note: Do the Battery Voltage Checks on pages 8 and 9 before doing Test 1.

Test 1 Do the following steps to test the alternator and the regulator-to-alternator cable:

Step A. With the engine off disconnect the regulator-to-alternator cable at the regulator.

Step B. Set the multimeter up to read ohms (RX1, X1). Zero the meter.

Step C. Touch the black probe to the large pin D and the red probe to large pin C of the regulator-to-alternator cable.

1. If the meter reads 55 to 75 ohms, do Step D.
2. If the meter does not read 55 to 75 ohms, do Test 4.

Step D. Now, reverse the probes. Touch the black probe to large pin C and the red probe to large pin D. (See Fig 5-2).
1. If the meter needle just barely moves or does not move at all, do Step E.
2. If the meter reads less than 1000 (1K ohms), do Test 4.

Step E. Touch the black probe to medium size pin B and the red probe to medium size pin E. (See Fig 5-2).
1. If you read 1-4 ohms, do Step F.
2. If you do not read 1-4 ohms, do Test 4.

Step F. Touch the black probe to pin B and touch the red probe to pin C.
1. If the needle does not move, do Step G.
2. If the needle moves, do Test 4.

Step G. Touch the black probe to the connector shell and touch the red probe to pins B, C, D and E. (See Fig 5-2)
1. If the needle does not move, do Step H.
2. If the needle moves, do Test 4.

Step H. Touch the black probe to ground touch the red probe to pins B, C, D and E. (See Fig 5-2).
1. If the needle does not move, do Test 2. The alternator and cable are okay.
2. If the needle moves, do Test 4.

Test 2 Do the following steps to check the regulator-to-battery cable:

Step A. Disconnect the regulator-to-battery cable at the regulator (be careful not to short the cable connector sockets to ground) (See Fig 5-1).

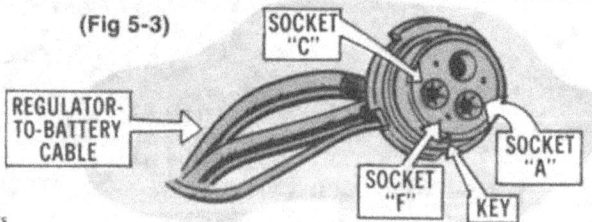

Step B. Set the multimeter up to read battery volts.

Step C. Ground the black probe to the vehicle frame and touch the red probe to the connector large socket A.

1. If you read 23-26 volts on the meter, do Step E.
2. If the meter does not read 23-26 volts, do Step D.

18

Step D. Clean the battery posts and terminals. Repeat Step C.

    1. If you now read 23-26 volts, repeat Check 5 on page 9.

    2. If you still do not read 23-26 volts, repair or replace the regulator-to-battery cable.

Step E. Be sure that the ignition switch is OFF.

Step F. Ground the black probe to the vehicle frame and touch the red probe to the cable connector small socket F.

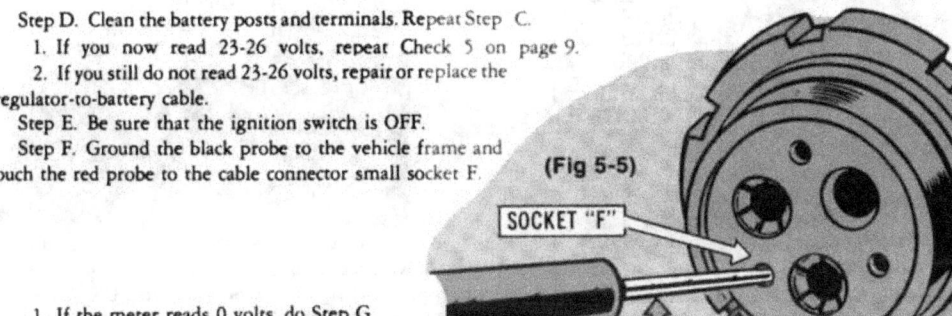

(Fig 5-5)

SOCKET "F"

RED PROBE

KEY

    1. If the meter reads 0 volts, do Step G.

    2. If the meter does not read 0 volts, do Test 5.

Step G. Turn the ignition switch ON.

Step H. Ground the black probe to the vehicle frame and touch the red probe to cable connector small socket F.

    1. If the meter reads 23-26 volts, do Step I.

    2. If the meter does not read 23-26 volts, do Test 5.

Step I. Turn the ignition switch OFF.

Step J. Set the multimeter up to read ohms (RX1, X1). "Zero" the meter.

Step K. Ground the black probe and touch the red probe to cable connector large socket C.

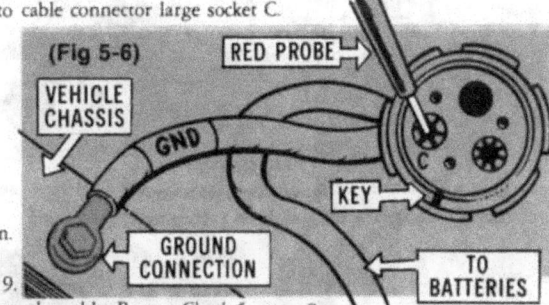

(Fig 5-6)

VEHICLE CHASSIS

RED PROBE

GND

KEY

C

GROUND CONNECTION

TO BATTERIES

    1. If the meter now reads 0 ohms, do Test 3.

    2. If the meter does not read 0 ohms, do Step L.

Step L. Clean and tighten cable pin C ground connection. Repeat Step K.

    1. If the meter now reads 0 ohms, repeat Check 5, page 9.

    2. If the meter still does not read 0 ohms, repair or replace the cable. Repeat Check 5, page 9.

Test 3 Do the following steps to adjust your regulator voltage output (if local SOP prohibits adjustment of the regulator at this level, replace the regulator and repeat Check 5, page 9):

Step A. Reconnect the connectors removed in the previous tests.

Step B. Remove the regulator adjustment access screw in the regulator (various regulators have printed or stamped locations of the screw).

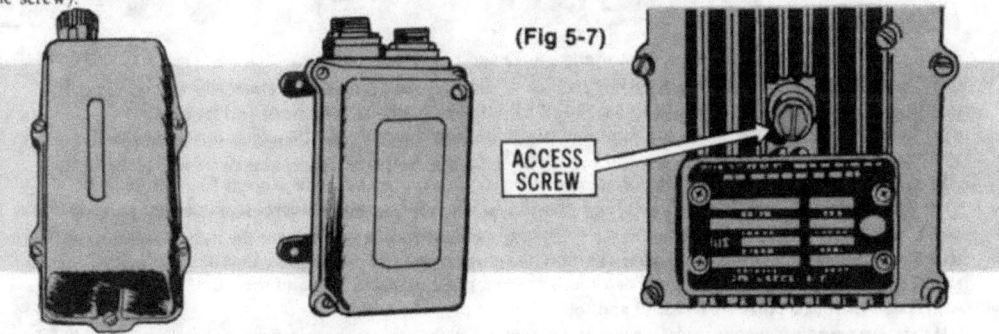

(Fig 5-7)

ACCESS SCREW

Step C. Start engine, turn on headlights and allow engine to reach normal operating temperature.

Step D. Increase engine RPM to 1000-1200 RPM after engine is warmed up.

Step E. Set the meter up to read battery volts.

Step F. Touch the black probe to the vehicle battery or slave receptacle negative terminal and touch the red probe to the positive terminal.

Public Domain, Google-digitized / http://www.hathitrust.org/access_use#pd-google

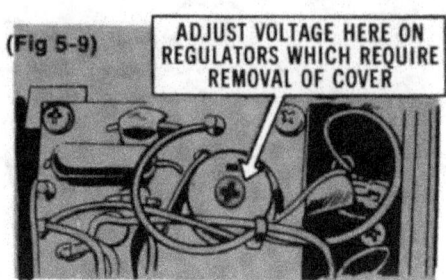

(Fig 5-8) TURN SCREW CLOCKWISE OR COUNTER CLOCKWISE

(Fig 5-9) ADJUST VOLTAGE HERE ON REGULATORS WHICH REQUIRE REMOVAL OF COVER

Step G. Attempt to adjust the regulator voltage to 27-29 volts by turning the voltage adjust screw clockwise or counter-clockwise. (See Fig 5-8 and 5-9 for adjusting through the access screw or with the cover off.)

  1. If the voltage will adjust and stabilize within the 27-29 volt range, the charging system now checks OK.

  2. If the voltage will not adjust to 27-29 volts, replace the regulator and repeat Check 5, page

Test 4 Do the following steps to test the regulator-to-alternator cable to determine if the cable is shorted or broken:

Step A. Disconnect the regulator-to-alternator cable at the alternator (regulator end already disconnected).

(Fig 5-10) DISCONNECT CABLE HERE — CABLE

Step B. Leave meter set as in previous test.

Step C. Touch the black and red probes to the following pins and sockets and observe the meter readings:

NOTE:
This cable is tied in with other vehicle cable harnesses. Bring the connector ends close enough together to make these tests.

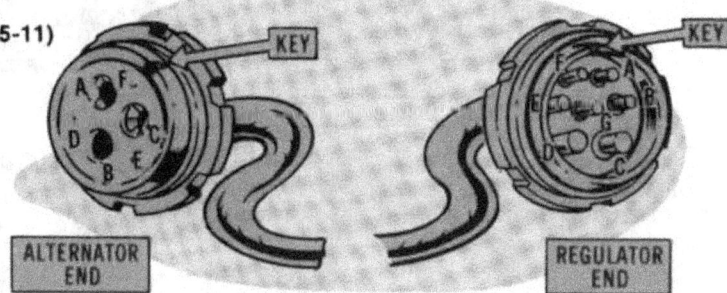

(Fig 5-11) KEY — KEY

ALTERNATOR END — REGULATOR END

  (a) Touch the black probe to the regulator end pin D and the red probe to the alternator end socket B.

  (b) Touch the black probe to regualtor end pin C and the red probe to alternator end socket C.

  (c) Touch the black probe to the regulator end pin B and red probe to the alternator end socket D.

  (d) Touch the black probe to the regulator end pin E and the red probe to the alternator end socket E.

  1. If the meter reads 0 for each of the above, do Step D.

  2. If the meter does not read 0 on any of the above, repair or replace the cable. Repeat Check 5, page 9.

Step D. Touch the black and red probes in the following socket sequence and observe the meter needle. (See Figure 5-11).

  (a) Touch the black probe to the connector shell, then touch the red probe to sockets B, C, D, and E.

  (b) Then touch the black probe to socket B and touch the red probe to sockets C, then D and E.

  (c) Touch the black probe to socket C and touch the red probe to sockets D and then E.

  (d) Touch the black probe to socket D and touch the red probe to socket E.

  1. If the meter needle moves or reads 0 on any of the tests above, the cable is shorted. Repair or replace the cable. Then repeat Check 5, page

  2. If the meter needle does not move, the cable is good. Replace the alternator. Repeat Check 5, page 9.

Public Domain, Google-digitized / http://www.hathitrust.org/access_use#pd-google

Test 5 Do the following steps to see if the ignition switch and its wiring are good:
Step A. Remove and drop the ignition switch down behind the dash panel.

(Fig 5-12)

NOTE: *Wire letters A, B, C, and D are stamped on the end of the rubber connectors. If you cannot see the letters, use the locator key in Figure 5-13. Wire numbers are tagged on the leads.*

(Fig 5-13)

(Fig 5-14)

Step B. Reconnect the switch handle. Disconnect wire #1/55.

Step C. Turn the ignition switch off.
Step D. Leave meter set up as is.
Step E. Touch black probe to a good ground and touch red probe to terminal D of the ignition switch (Terminal D is where you removed Wire #1).

(Fig 5-15)

1. If you get a voltage reading, replace the switch and repeat Check 5, on page
2. If you get no voltage reading, do Step F.

Step F. Turn the ignition switch ON.
1. If you get no voltage reading, turn the switch OFF and do Test 6.
2. If the meter reads battery voltage (23-26 volts), Wire #1 is bad. Direct Support must repair or replace the wiring harness.

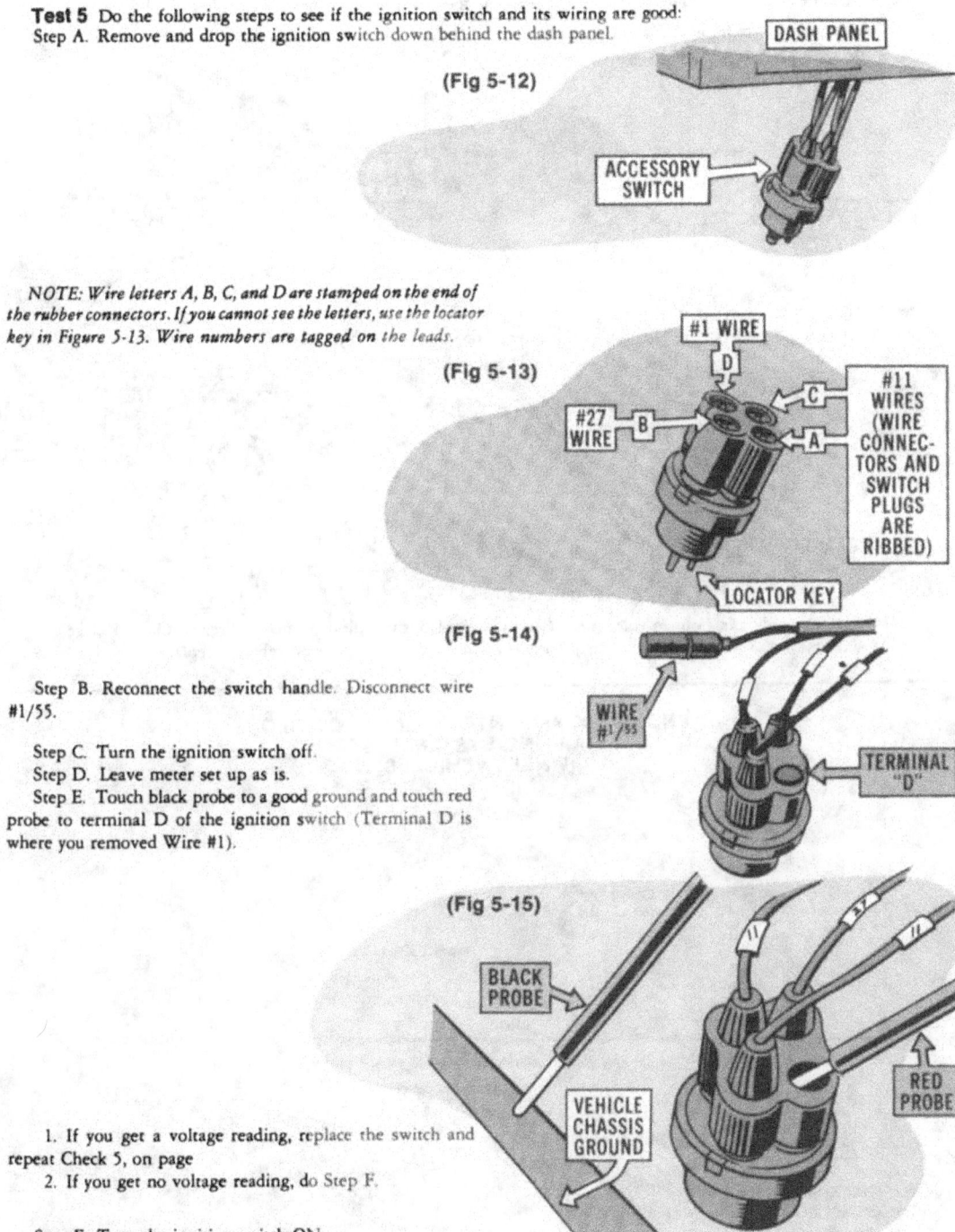

21

Test 6- Do the following steps to test the wiring between the ignition switch and the batteries.

Step A. Reconnect Wire #1 to the ignition switch.

Step B. Disconnect both #11 wires from the switch.   (Fig 5-16)

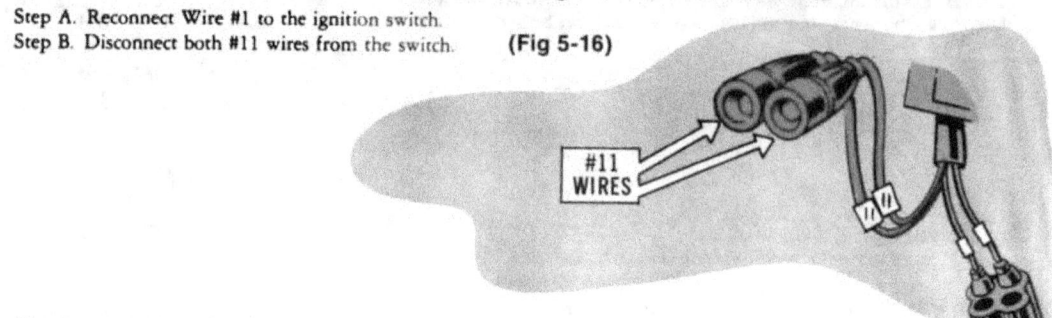

Step C. Leave meter set up as is.

Step D. Touch black probe to a good ground and touch red probe, in turn, to the end of each #11 wire.

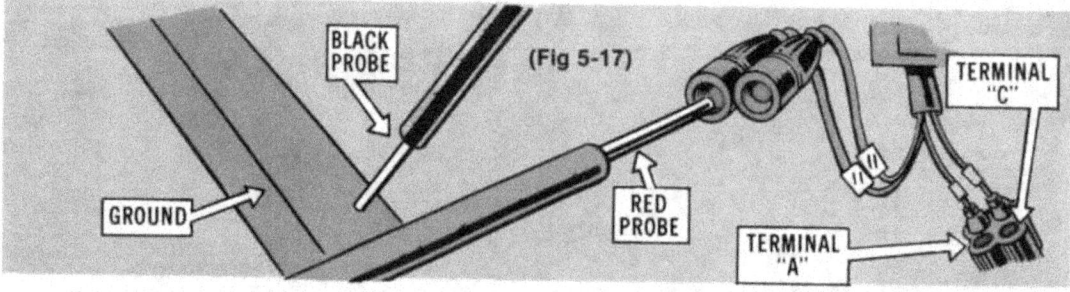

1. If the needle reading is 23 to 26 volts at each #11 wire, replace the ignition switch and repeat Check 5, page

2. If you do not get a voltage reading at one or both #11 wires, have Direct Support repair or replace the wiring harness.

**END OF 100 AMP INTERNALLY RECTIFIED
CHARGING SYSTEM TESTS
(WHEEL VEHICLES)**

Chapter 6

100 AMP, EXTERNALLY RECTIFIED, PLATE TYPE
(WHEEL)

NOTE: Do the Battery Voltage Check on pages 8 and 9 before doing Test 1.

Test 1- Do the following steps to test the rectifier-to- alternator cable and the alternator:

Step A. With engine off, disconnect the rectifier-to- alternator cable at the rectifier.

(Fig 6-1)

Step B. Set the multimeter up to read ohms (RX1, X1). Zero the meter.

(Fig 6-2)

Step C. Touch the black probe to large pin A and the red probe to large pin B and then pin C.
1. If you read 0 ohms, do Step D.
2. If you do not read 0 ohms, do Test 5.

Step D. Touch the black probe to small pin D and touch the red probe to small pin E. (See Fig 6-2).
1. If you read 1-4 ohms, do Step E.
2. If you do not read 1-4 ohms, do Test 5.

Step E. Touch the black probe to pin B and touch the red probe to pin C. (See Fig 6-2).
1. If the needle does not move, do Step F.
2. If the needle moves, do Test 5.

Step F. Touch the black probe to the connector shell and touch the red probe to pin A, B, C, D and E. (See Fig 6-2).
1. If the needle does not move, do Step G.
2. If the needle moves, do Test 5.

Step G. Touch the black probe to ground, touch the red probe to pins A, B, C, D and E. (See Fig 6-2).
1. If the needle does not move, do Test 2.
2. If the needle moves, do Test 5.

Test 2- Do the following steps to test the regulator-to- rectifier cable and the rectifier:

Step A. Be sure to hook the rectifier-to-alternator cable back up.

Step B. Disconnect the rectifier-to-regulator cable at the regulator.

(Fig 6-3)

Step C. Set multimeter up to read ohms (RX1, X1). Zero the meter.

Step D. Touch the black probe to the large pin D and the red probe to large pin D and the red probe to large pin C. (See Fig 6 -4).
1. If the needle reads 55 to 75 ohms, do Step E.
2. If the needle does not read 55 to 75 ohms, do Test 6.

23

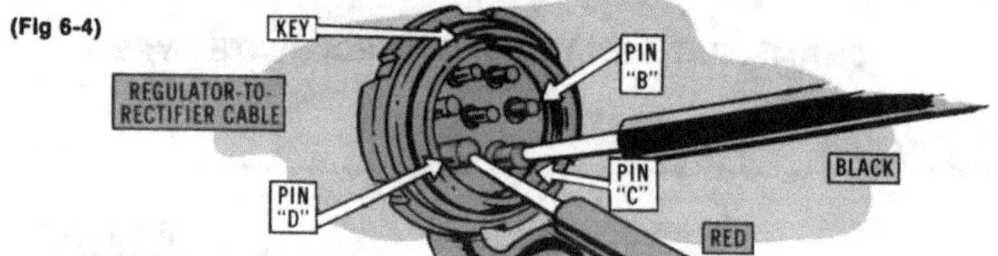

(Fig 6-4)

KEY | REGULATOR-TO-RECTIFIER CABLE | PIN "B" | PIN "C" | BLACK | RED | PIN "D"

Step E. Now, reverse the probes. Touch the black probe to large pin C and the red probe to large pin D.

1. If the needle just barely moves or does not move at all, do Step F.

2. If the needle reads less than 1000 (1K ohms), do Test 6 (See Fig 6-4).

Step F. Touch the black probe to pin B and red probe to pin E. (See Fig 6-4).

1. If the meter reads 1-4 ohms, do Step G.

2. If the meter does not read 1-4 ohms, do Test 6.

Step G. Touch the black probe to pin B and the red probe to pin C, then pin D. (See Fig 6-4).

1. If the needle does not move, do Step H.

2. If the needle moves, do Test 6.

Step H. Touch the black probe to the cable connector shell and touch the red probe to cable pin B, C and D. Then, touch the black probe to ground and touch the red to pins B, C and D. (See Fig 6-4).

1. If the needle does not move, do Test 3.

2. If the needle moves, do Test 6.

Test 3 Do the following steps to check the regulator-to-battery cable:

Step A. Disconnect the regulator-to-battery cable at the regulator.

(Fig 6-5)

REGULATOR-TO-BATTERY CABLE | KEY

Step B. Set the multimeter up to read battery volts.

(Fig 6-6)

SOCKET "A" | RED | VEHICLE FRAME | BLACK | KEY

Step C. Ground the black probe to the vehicle frame and touch the red probe to the cable connector large socket A.

1. If you read 23-26 volts on the meter, do Step E.

2. If the meter reading is 0 volts, do Step D.

Step D. Clean the battery posts and terminals. Repeat Step C.

1. If the voltage now reads 23-26 volts, repeat Check 5, page 9.

2. If the meter reading is still 0 volts, get DS to repair or replace the regulator-to-battery cable. Repeat Check 5, page 9.

Step E. Be sure that the ignition switch is OFF.

Step F. Ground the black probe to the vehicle frame and touch the red probe to the cable connector small socket F. (See Fig 6-6).

(Fig 6-7)

BLACK | SOCKET "F" | KEY | RED

24

1. If the meter reads 0 volts, do Step G.
2. If the meter does not read 0 volts, do Test 7.
Step G. Turn the ignition switch ON.
Step H. Ground the black probe to the vehicle frame and touch the red probe to the cable connector small socket F. (See Fig 6-6).
  1. If the meter reads 23-26 volts, do Step I.
  2. If the meter reads 0 volts, do Test 7.

Step I. Turn the ignition switch OFF.
Step J. Set the multimeter up to read ohms (RX1, X1). Zero the meter.
Step K. Ground the black probe and touch the red probe to cable connector large socket C. (See Fig 6-7).
  1. If the meter reads 0 ohms, do Test 4.
  2. If the meter does not read 0 ohms, do Step L.

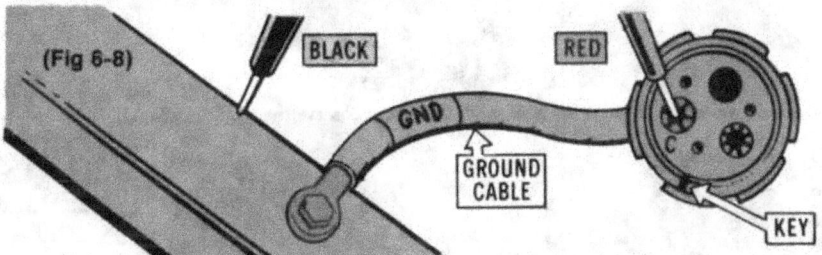

(Fig 6-8)

BLACK

RED

GND

GROUND CABLE

KEY

Step L. Clean and tighten the cable pin C ground connection. Repeat Step K.
  1. If the meter now reads 0, do Test 4.

2. If the meter still does not read 0 ohms, replace or repair the cable. Repeat Check 5, page 9.

Test 4 Do the following steps to adjust the regulator voltage output (if local SOP prohibits adjustment of the regulator at this level, replace the regulator and repeat Check 5, page 9):
Step A. Reconnect the connectors and connections removed in the previous tests.
Step B. Remove the regulator adjustment access screw or the cover to expose the voltage adjustment screw in the regulator.

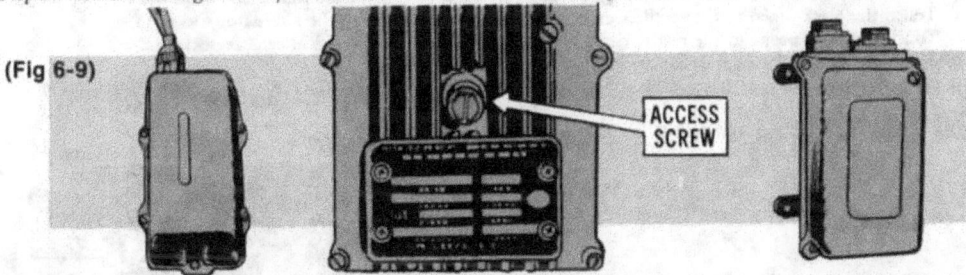

(Fig 6-9)

ACCESS SCREW

Step C. Start engine (1000 to 1200 RPM); turn on high beam lights and allow the engine to reach normal operating temperature.
Step D. Set the multimeter up to read battery volts.
Step E. Connect the black probe to the vehicle battery negative terminal and connect the red probe to the positive terminal.
Step F. Attempt to adjust the regulator voltage to 27-29 volts by turning the voltage adjust screw clockwise or counter-clockwise.

(Fig 6-10)

TURN SCREW CLOCKWISE OR COUNTER-CLOCKWISE

ADJUST VOLTAGE HERE ON REGULATORS WHICH REQUIRE REMOVAL OF COVER

(Fig 6-11)

ABOVE SCREW TO ADJUST VOLTAGE
ULATOR ENGINE GENERATOR

1. If the voltage will adjust and stabilize within the 27-29 volt range, the charging system now checks OK.
2. If the voltage will not adjust to 27-29 volts, replace the regulator. Repeat Check 5, page 9.

Test 5 Do the following steps to test the rectifier-to-alternator cable to determine if the cable is shorted or broken:
Step A. Disconnect the rectifier-to-alternator cable at the alternator (rectifier end already disconnected).

(Fig 6-12)

Step B. Leave meter set as in previous test.
Step C. Touch the black and red probes to the following pins and sockets and observe the meter readings:

(Fig 6-13)

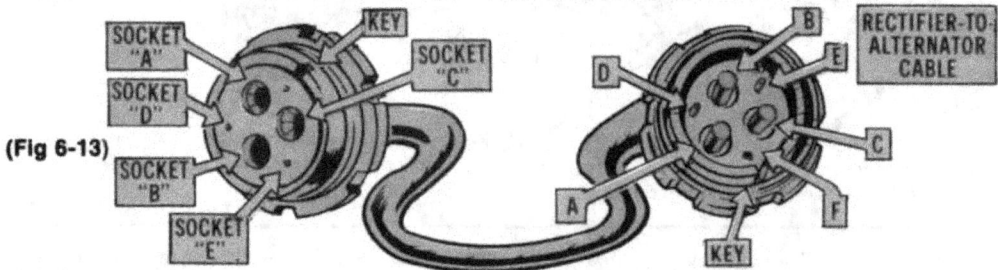

(a) Touch the black probe to rectifier end pin A and the red probe to alternator end socket A.
(b) Touch the black probe to the rectifier end pin B and red probe to the alternator end socket B.
(c) Touch the black probe to rectifier end pin C and the red probe to the alternator end socket C.
(d) Touch the black probe to the rectifier end pin D and the red probe to the alternator socket D.
(e) Touch the black probe to the rectifier end pin E and the red probe to the alternator socket E.
1. If the meter reads 0 for each of the above, do Step D.
2. If the meter does not read 0 ohms on any of the above, repair or replace the cable. Repeat Check 5, page 9.

Step D. Touch the black and red probes in the following pin sequence and observe the meter needle:

(Fig 6-14)

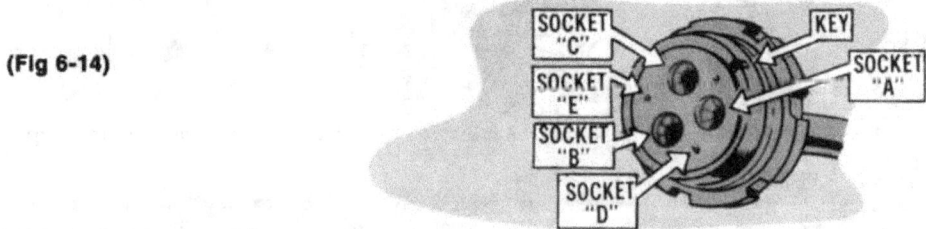

(a) Touch the black probe to the connector shell, then touch the red probe to sockets A, B, C, D and E.
Then, touch the black probe to ground and touch the red probe to sockets A, B, C, D and E.
(b) Touch the black probe to socket A and touch the red probe to sockets, B, C, D and E.
(c) Touch the black probe to socket B and touch the red probe to sockets C, then D and E.
(d) Touch the black probe to socket C and touch the red probe to sockets D then E.
(e) Then touch the black probe to socket D and touch the red probe to socket E.
1. If the meter needle moves on any of the preceding steps the cable is shorted. Repair or replace the cable. Repeat Check 5, page 9.
2. If the meter needle does not move, the cable is good. Replace the alternator. Then repeat Check 5, page 9.

Test 6- Do the following steps to test the regulator-to-rectifier cable:
Step A. Disconnect the regulator-to-rectifier cable at the regulator (rectifier end already disconnected).

(Fig 6-15)

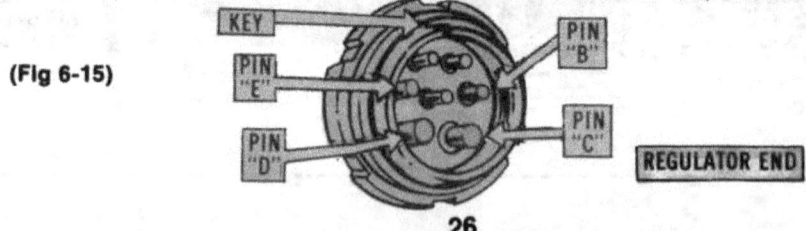

26

Step B. Set the meter to read ohms (RX1, X1). Zero the meter.

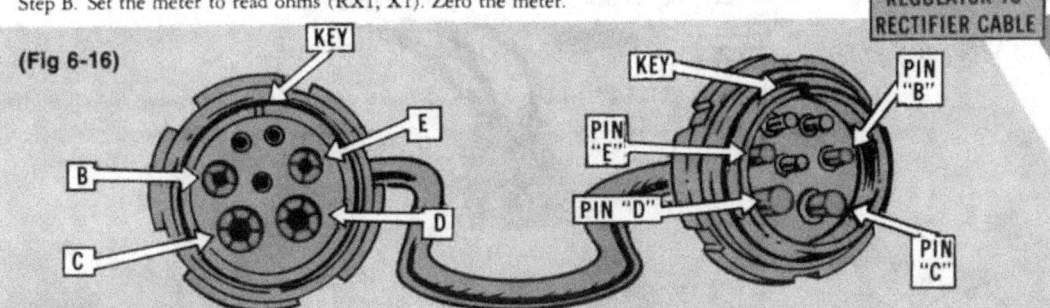

(Fig 6-16)

KEY

E

B

C

D

KEY

PIN "E"

PIN "D"

PIN "B"

PIN "C"

Step C. Connect the black and red probes in the following pin and socket sequence and observe the meter needle.
1. Touch black probe to pin B and red probe to socket B.
2. Touch black probe to pin C and red probe to socket C.

3. Touch black probe to pin D and red probe to socket D.
4. Touch black probe to pin E and red probe to socket E.
   (a) If the meter needle reads "0" ohms, do Step D.
   (b) If the meter needle does not read 0 ohms, repair or replace the cable. Then repeat Check 5, page 9.

Step D. Touch the black and red probes in the following sockets and observe the meter needle:
1. Touch the black probe to the connector shell, then touch the red probe in turn, to sockets B, C, D and E. Then, touch the black probe to ground and the red probe to sockets B, C, D and E.

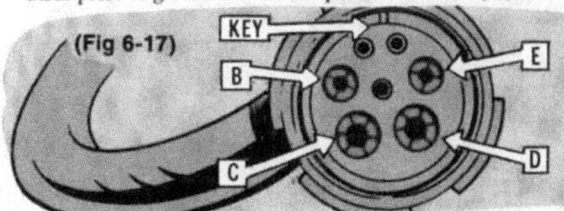

(Fig 6-17)

KEY

B

E

C

D

2. Now, touch the black probe to socket B and the red probe, in turn, to sockets C, D and E. Touch the black probe to socket C and the red probe to sockets D and E. Touch the black probe to socket D and the red probe to socket E.
   (a) If the meter needle moves, repair or replace the rectifier-to-regulator cable. Repeat Check 5, page 9.
   (b) If the meter needle does not move, replace the rectifier. Repeat Check 5, page 9.

Test 7- To test to see if the ignition switch and its wiring are good, to the following steps:

Step A. Remove and drop the ignition switch down behind the dash panel.

(Fig 6-18)

DASH PANEL

IGNITION SWITCH

NOTE: *Wire letters A, B, C and D are stamped on the end of the rubber connectors. If you can not see the letters, use the locator key in (Fig 6-19). Wire numbers are tagged on the leads.*

(Fig 6-19)

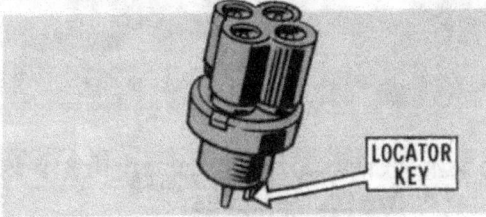

LOCATOR KEY

27

Step B. Reconnect the switch handle. Disconnect Wire #1 from the switch (Fig 6-20)

(Fig 6-20)

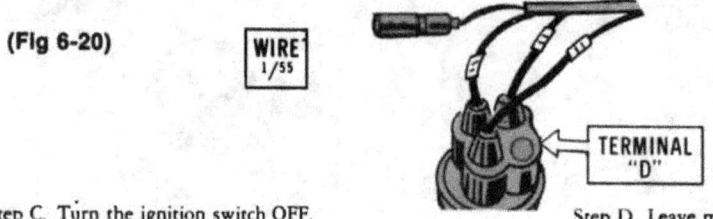

Step C. Turn the ignition switch OFF.

Step D. Leave meter set up as is.

(Fig 6-21)

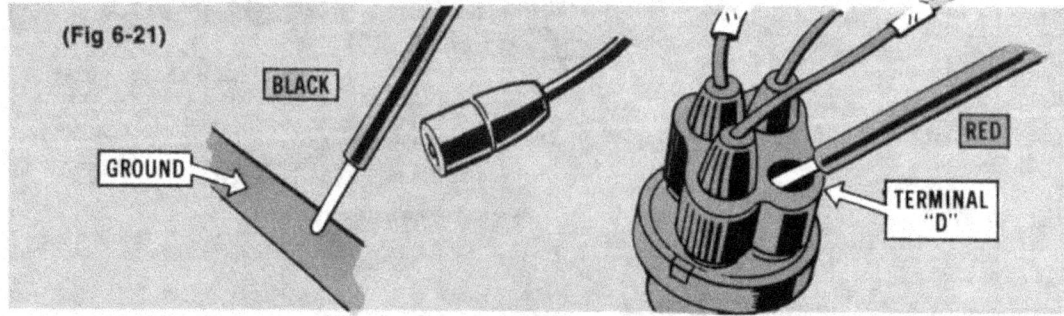

Step E. Touch black probe to a good ground and touch red probe to terminal D of the ignition switch (Terminal D is where you removed Wire #1).

1. If you get a voltage reading, replace the switch. Repeat Check 5, page 9.
2. If you get no voltage reading, do Step F.

Step F. Turn the ignition switch ON.

1. If you get no voltage reading, turn the switch OFF and do Test 8.
2. If the meter reads battery voltage (23-26 volts), Wire #1 is bad. Direct support must repair or replace the wiring harness.

Test 8- To test to see if the wiring between the ignition switch and the batteries is good, do the following steps:

(Fig 6-22)

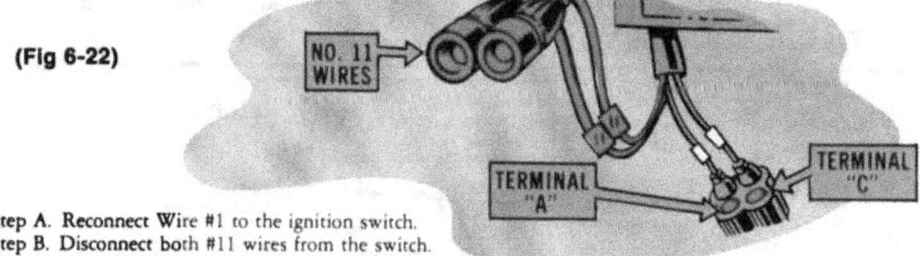

Step A. Reconnect Wire #1 to the ignition switch.
Step B. Disconnect both #11 wires from the switch.
Step C. Leave meter set up as is.
Step D. Touch black probe to a good ground and touch red probe, in turn, to the end of each #11 wire.

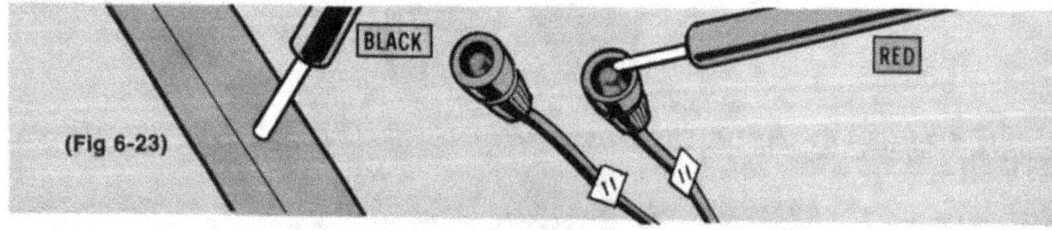

(Fig 6-23)

1. If the needle reading is 23 to 26 volts at each #11 wire, replace the ignition switch and repeat Check 5, page 9.
2. If you do not get a voltage reading at one or both #11 wires, have Direct Support repair or replace the wiring harness.

**END OF 100 AMP, EXTERNALLY RECTIFIED, PLATE TYPE
CHARGING SYSTEMS TESTS
(WHEEL VEHICLES)**

Chapter 7

100 AMP INTERNALLY RECTIFIED
(TRACK)

NOTE: Do the Battery Voltage Checks on page 8 and 9 before doing Test 1.

Test 1- Do the following steps to test the alternator and the regulator-to-alternator cable:

Step A. With the engine off disconnect the regulator-to-alternator cable at the regulator.

(Fig 7-1)

Step B. Set the multimeter up to read ohms (RX1, X1). "Zero" the meter.

Step C. Connect the black probe to large pin D and red probe to large pin C of the cable.

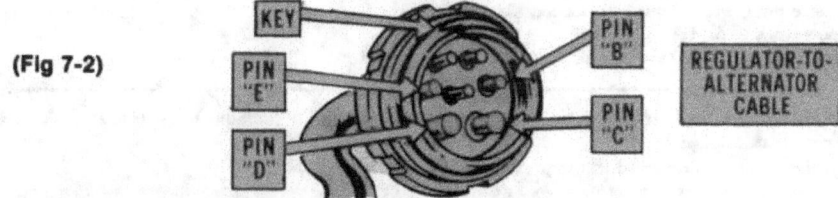

(Fig 7-2)

1. If the meter reads 55 to 75 ohms, do Step D.
2. If the meter does not read 55 to 75 ohms, do Test 4, Steps A, B and C.

Step D. Now, connect the black probe to large pin C and the red probe to large pin D.) (See Fig 7-2).
1. If the meter needle just barely moves or does not move at all, do Step E.
2. If the meter reads less than 1000 (1K ohms), do Test 4, Steps A, B and D.

Step E. Connect the black probe to medium pin B and the red probe to medium pin E.
1. If the meter reads 1-4 ohms, do Step F.
2. If the meter does not read 1-4 ohms, do Test 4, Steps A, B and E.

Step F. Touch the black probe to medium pin B and touch the red probe to large pins C, D and E.
1. If the needle does not move, do Step G.
2. If the needle moves or reads 0 ohms, do Test 4, Steps A, B and F.

Step G. Touch the black probe to ground, and touch the red probe to pins B, C, D and E.
1. If the needle does not move, do Step H.
2. If the needle moves, do Test 4, Steps A, B and C.

Step H. Touch the black probe to the connector shell and touch the red probe to pins B, C and D. Then, touch the black probe to ground and the red probe to pins B, C and D. (See Fig 7-2).
1. If the needle does not move, do Test 2.
2. If the needle moves or reads 0 ohms, the wires in the connector are shorted. Repair or replace the wires.

Test 2- Do the following steps to check the regulator-to-battery cable:

(Fig 7-3)

Step A. Disconnect the regulator-to-battery cable at the regulator.

Step B. Set the multimeter up to read battery volts. Turn on the master switch.

Step C. Ground the black probe to the vehicle frame and touch the red probe to the cable connector large socket A.

(Fig 7-4)

1. If the meter reads 23-26 volts, do Step E.
2. If the meter reading is less than 23 volts, repair or replace the regulator-to-battery cable.

Step D. Start engine and run at 1000-1200 RPM.

Step E. Ground the black probe and touch the red probe to the cable, small socket F. (See Fig 7-4).
1. If the meter reads less than 23 volts, do Step F.
2. If the meter reads 23-26 volts, do Step F.

Step F. Turn the master switch OFF.

Step G. Set the multimeter up to read ohms (RX1, X1). Zero the meter.

Step H. Ground the black probe and touch the red probe to cable connector large socket C. (See Fig 7-4).
1. If the meter reads 0 ohms, do Test 3.
2. If the meter does not read 0 ohms, do Step I.

Step I. Clean the cable pin C ground wire connection. Repeat Step H.
1. If the meter now reads 0, do Test 3.
2. If the meter still does not read 0 ohms, repair or replace the cable. Repeat Check 5, page 9.

Test 3- Do the following steps to adjust your regulator voltage output (if local SOP prohibits adjustment at this level, replace the regulator and repeat Check 5, page 9):

Step A. Reconnect the connectors removed in the previous test.

Step B. Remove the regulator adjustment access screw or the cover to expose the voltage adjustment screw in the regulator.

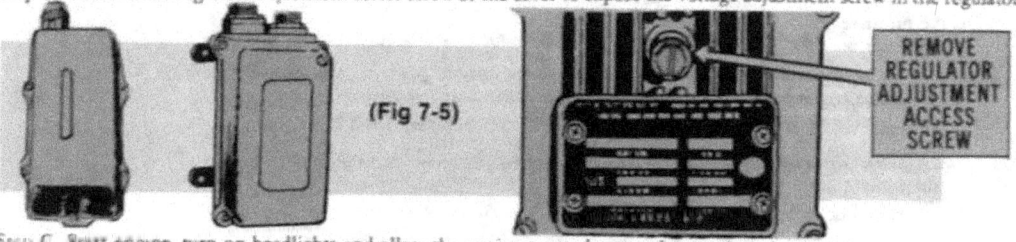

(Fig 7-5)

Step C. Start engine, turn on headlights and allow the engine to reach normal operating temperature.

Step D. Increase engine speed to 1000 to 1200 RPM after engine is warmed up.

Step E. Set multimeter up to read battery volts.

Step F. Touch the black probe to the vehicle battery or slave receptacle negative terminal and touch the red probe to the positive terminal.

Step G. Attempt to adjust the regulator voltage to 27-29 volts by turning the voltage adjust screw clockwise or counter clockwise.

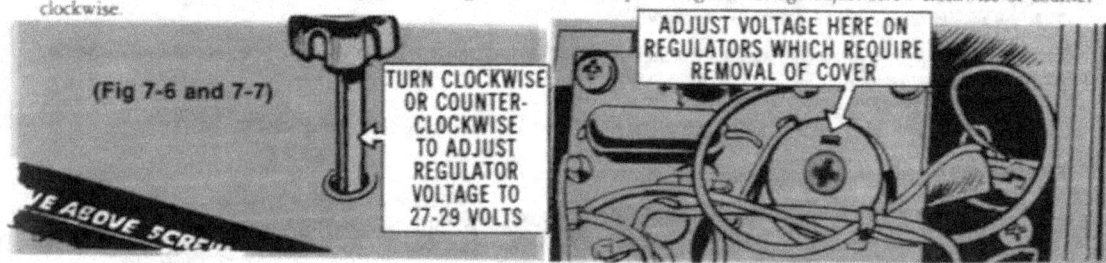

(Fig 7-6 and 7-7)

1. If the voltage adjusts and holds steady within the 27-29 volts range, the charging system now checks OK.
2. If the voltage will not adjust to 27-29 volt, replace the regulator and repeat Check 5, page 9.

Test 4 - Do the following steps to test the bulkhead-to-alternator cable and the alternator:

Step A. With the engine off, disconnect the bulkhead-to-alternator cable at the bulkhead.

30

(Fig 7-8)

DISCONNECT
BULKHEAD
CABLE

Step B. Set the multimeter up to read ohms.
Step C. Touch the black probe to the large pin D and the red probe to large pin C of the Cable.

(Fig 7-9)

KEY

PIN
"D"

BLACK

RED

PIN
"C"

BULKHEAD TO
ALTERNATOR
CABLE

   1. If the meter reads 55 to 75 ohms, repair or replace the regulator-to-bulkhead cable, repeat Check 5, page 9.
   2. If the meter does not read 55 to 75 ohms, do Test 5.
Step D. Now reverse the probes. Touch the black probe to large pin C and the red probe to large pin D. (See Fig 7-9).
   1. If the meter needle just barely moves or does not move at all, repair or replace the regulator-to-bulkhead cable. Repeat Check 5, page 9.
   2. If the meter reads less than 1000 (1K ohms) do Test 5.
Step E. Connect the black probe to medium size pin B and the red probe to medium size pin E. (See Fig 7-9).
   1. If the meter reads 1-4 ohms, repair or replace the regulator-to-bulkhead cable. Repeat Check 5, page 9.
   2. If the meter does not read 1-4 ohms, do Test 5.
Step F. Touch the black probe to pin B and touch the red probe to pins C, D and E. (See Fig 7-9)
   1. If the needle does not move, repair or replace the regulator-to-bulkhead cable. Repeat Check 5, page 9.
   2. If the needle moves, do Test 5.
Step G. Touch the black probe to ground, and touch the red probe to pins B, C, D and E.
   1. If the needle now does not move, repair the short to ground on the cable between the bulkhead and the regulator.
   2. If the needle moves, do Test 5.
Step H. Touch the black probe to the connector shell and touch the red probe to pins B, C and D. Then, touch the black probe to ground and touch the red probe to pins B, C and D. (See Fig 7-9).
   1. If the needle does not move, repair or replace the regulator-to-bulkhead cable.
   2. If the needle moves or reads 0 ohms, the wires of the connector are shorted. Repair or replace the wires.

Test 5- Do the following steps to test the bulkhead-to-alternator cable to determine if the cable is shorted or broken:
   Step A. Disconnect the bulkhead-to-alternator cable at the alternator (bulkhead end already disconnected).

(Fig 7-10)

ALTERNATOR

DISCONNECT
BULKHEAD
TO REGULATOR
CABLE

   Step B. Set the meter to read ohms (RX1, X1). "Zero" the meter.
   Step C. Touch the black and red probes to the following pins and sockets and observe the meter readings.

(Fig 7-11)

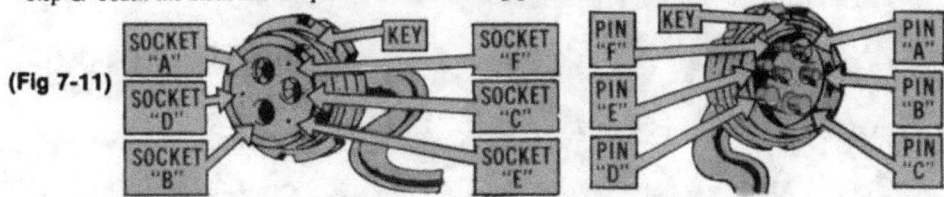

SOCKET
"A"

KEY

SOCKET
"F"

KEY

PIN
"F"

PIN
"A"

SOCKET
"D"

SOCKET
"C"

PIN
"E"

PIN
"B"

SOCKET
"B"

SOCKET
"E"

PIN
"D"

PIN
"C"

   (a) Touch the black probe to bulkhead end pin C and the red probe to alternator end socket C.

(b) Touch the black probe to bulkhead end pin B and red probe to alternator end socket D.

(c) Touch the black probe to pin E and the red probe to socket E.

(d) Touch the black probe to bulkhead end pin D and the red probe to alternator end socket B.

1. If the meter reads 0 for each of the above, do Step D.

2. If the meter needle does not move on any of the above, repair or replace the cable. Repeat Check 5, page 9.

Step D. Touch the black and red probes in the following socket sequence and observe the meter needle:

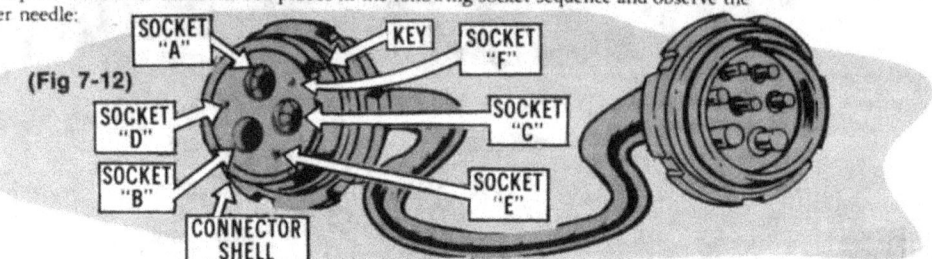

(Fig 7-12)

(a) Touch the black probe to the connector shell, then touch the red probe to sockets B, C, D and E. Touch the black probe to ground and the red probe to sockets B, C, D and E.

(b) Then, touch the black probe to socket B and touch the red probe to sockets C, D and E.

(c) Touch the black probe to socket C and touch the red probe to sockets D and E.

(d) Touch the black probe to socket D and touch the red probe to socket E.

1. If the meter needle moves or reads 0 on any of the tests above, the alternator is good. Repair or replace the cable. Then repeat Check 5, page 9.

2. If the meter needle does not move, the cable is good. Replace the alternator. Repeat Check 5, page 9.

Test 6 With the engine off do the following steps to test the generator field switch:

(Fig 7-13)

Step A. Remove the connector from the generator field switch. Use a short jumper wire. Short the contacts of sockets A and B in the field switch cable plug.

(Fig 7-15)

Step B. Set multimeter to read battery volts.

Step C. Ground the black probe and touch the red probe to the regulator-to-battery cable socket F.

(Fig 7-15)

32

1. If the meter reads 23-26 volts, replace the generator field switch, repeat Check 5, page 9.

2. If the meter reads 0 volts, do Step D.

Step D. At the generator field switch plug, ground the black probe to the vehicle and touch the red probe to cable socket A.

(Fig 7-16)

1. If you read 0 volts, repair or replace cable 1A. Repeat Check 5, page 9.

2. If you read 23-26 volts, do Step E.

Step E. Set multimeter to read ohms, (RX1, X1). Zero the meter.

Step F. Start engine and run at 1000-1200 RPM.

Step G. Touch the black probe to pin A and the red probe to pin B of the generator field switch.

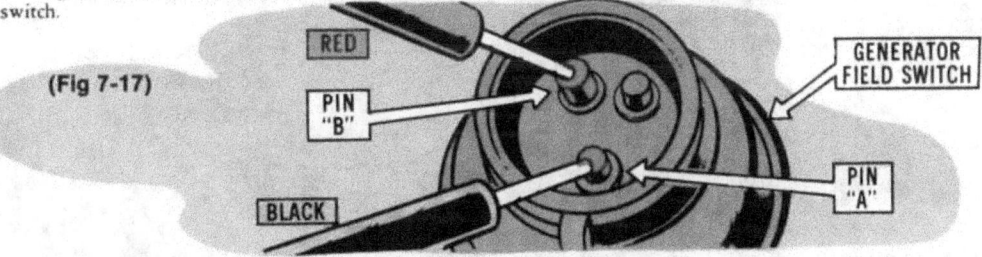

(Fig 7-17)

1. If you read 0 ohms, repair or replace wire #1B between the field switch and the regulator. Repeat Check 5, page 9.

2. If you do not read 0 ohms, replace the generator field switch. Repeat Check 5, page 9.

END OF 100 AMP INTERNALLY
RECTIFIED CHARGING SYSTEM TESTS
(TRACK)

Chapter 8
100 AMP, EXTERNALLY RECTIFIED, PLATE TYPE
(TRACK)

NOTE: Do the Battery Voltage Checks on pages 8 and 9 before doing Test 1.

Test 1- Do the following steps to test the rectifier-to- alternator cable and the alternator:

Step A. With engine off, disconnect the rectifier-to- alternator cable at the rectifier.

(Fig 8-1)

Step B. Set the multimeter up to read ohms (RX1, X1). "Zero" the meter.

(Fig 8-2)

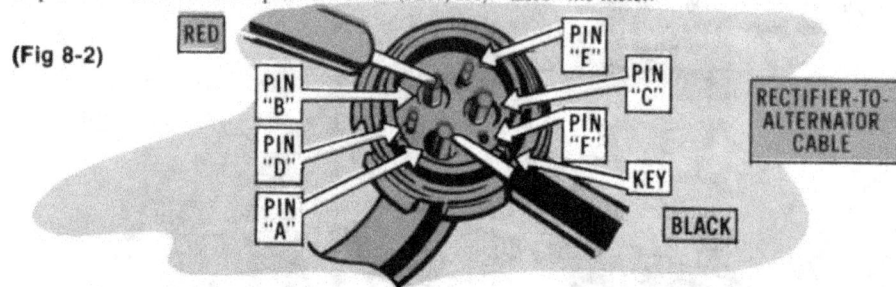

Step C. Touch the black probe to large pin A and the red probe to large pin B and then pin C.
  1. If you read 0 ohms, do Step D.
  2. If you do not read 0 ohms, do Test 5.

Step D. Touch the black probe to small pin D and touch the red probe to small pin E. (See Fig 8-2).
  1. If you read 1-4 ohms, do Step E.
  2. If you do not read 1-4 ohms, do Test 5.

Step E. Touch the black probe to large pin C and touch the red probe to small pin E. (See Fig 8-2).
  1. If the needle does not move, do Step F.
  2. If the needle does move, do Test 5.

Step F. Touch the black probe to ground, and touch the red probe to pins A, B, C, D and E (See Fig 8-2).
  1. If the needle does not move, do Step G.
  2. If the needle moves, do Test 5.

Step G. Touch the black probe to the connector shell, and touch the red probe to pins A, B, C, D and E.
  1. If the needle does not move, do Test 2.
  2. If the needle moves, the wires in the connector shell are shorted. Repair or replace the wires.

Test 2- Do the following steps to test the regulator-to-rectifier cable and the rectifier:

Step A. Reconnect the rectifier-to-alternator cable at the rectifier.

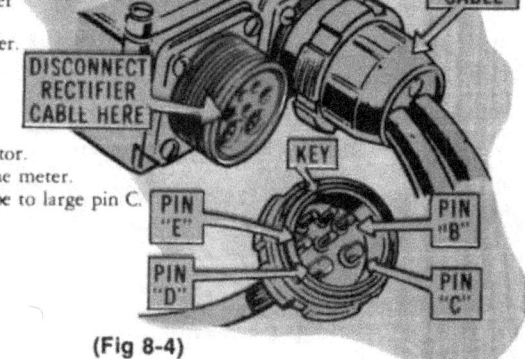

(Fig 8-3)

Step B. Disconnect the regulator-to-rectifier cable at the regulator.
Step C. Set the multimeter to read ohms (RX1, X1). "Zero" the meter.
Step D. Touch the black probe to large pin D and the red probe to large pin C.

(Fig 8-4)

1. If the meter reads 55 to 75 ohms, do Step E.
2. If the meter does not read 55-75 ohms, do Test 6, Steps A and B.

Step E. Touch the black probe to lare pin C and red probe to large pin D. (See Fig 8-4).
1. If the needle does not move, do Step F.
2. If the needle moves, do Test 6, Steps A and C.

Step F. Touch the black probe to medium pin B and the red probe to medium pin E. (See Fig 8-4).
1. If the meter reads 1-4 ohms, do Step G.
2. If the meter does not read 1-4 ohms, do Test 6, Steps A and D.

Step G. Touch the black probe to medium pin B and the red probe to large pin C and then D. (See Fig 8-4).
1. If the needle does not move, do Step H.
2. If the needle moves, do Test 6, Steps, A, and E.

Step H. Then, touch the black probe to ground and touch the red probe to pins B, C, D and E.
1. If the needle does not move, do Step I.
2. If the needle moves, do Test 6, Steps A and F.

Step I. Touch the black probe to the cable connector shell and the red probe to pin B, C, D and E. (See Fig 8-4).
1. If the needle moves, the wires in the connector are shorted. Repair or replace the wires.
2. If the needle does not move, do Test 3.

Test 3- Do the following steps to check the regulator-to-battery cable:
Step A. Disconnect the regulator-to-battery cable at the regulator.

(Fig 8-5)

Step B. Set the multimeter up to read battery volts. Turn on the master switch.
Step C. Ground the black probe to the vehicle frame and touch the red probe to the cable connector large socket A.

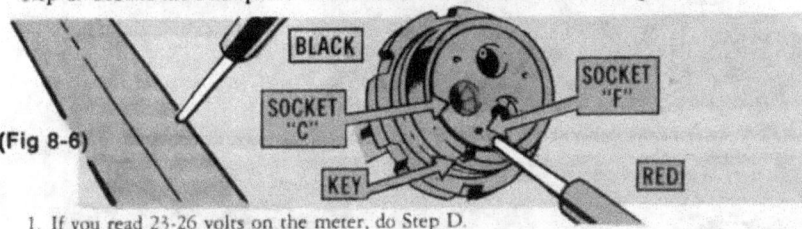
(Fig 8-6)

1. If you read 23-26 volts on the meter, do Step D.
2. If the meter reading is 0 volts, repair or replace the cable between the regulator and the master relay.

Step D. Start engine and run at 1000-1200 RPM.
Step E. Ground the black probe and touch the red probe to small socket F. (See Fig 8-6).
1. If the meter reads 0 volts, do Test 8.
2. If the meter reads 23-26 volts, do Step F.

Step F. Turn the MASTER switch OFF.
Step G. Set the multimeter up to read ohms.
Step H. Ground the black probe and touch the red probe to cable connector large socket C. (See Fig 8-6).
1. If the meter reads 0 ohms, do Test 4.
2. If the meter does not read 0 ohms, do Step I.

Step I. Clean the cable pin C ground wire connection. Repeat Step H.
1. If the meter now reads 0, do Test 4.
2. If the meter does not read 0 ohms, replace the cable. Repeat Check 5, page 9.

Test 4- Do the following steps to adjust your regulator voltage output (if local SOP prohibits adjustment of the regulator at this level, replace the regulator and repeat Check 5, page 9):
Step A. Reconnect the connectors removed in the previous tests.
Step B. Remove the regulator adjustment access screw or the cover to expose the voltage adjustment screw in the regulator.

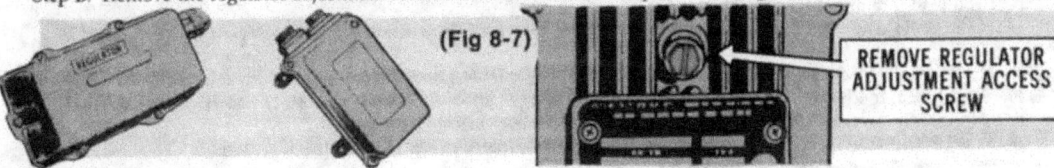
(Fig 8-7) REMOVE REGULATOR ADJUSTMENT ACCESS SCREW

Step C. Start engine, turn on highbeam headlights and allow to reach normal operating temperature.
Step D. After engine is warmed up, increase RPM to 1000 to 1200.
Step E. Set multimeter up to read battery volts.
Step F. Connect the black probe to the vehicle battery negative terminal and connect the red probe to the positive terminal.

(Fig 8-8) TURN CLOCKWISE OR COUNTER-CLOCKWISE

(Fig 8-9) VOLTAGE ADJUSTMENT

Step G. Attempt to adjust the regulator voltage to 27-29 volts by turning the voltage adjust screw clockwise or counter clockwise.
   1. If the voltage will adjust and stabilize within the 27-29 volt range the charging system now checks OK.
   2. If the voltage will not adjust to 27-29 volt, replace the regulator and repeat Check 5, page 9.

Test 5- Do the following steps to test the rectifier-to-alternator cable for continuity or shorts:
Step A. Set the multimeter to read ohms.

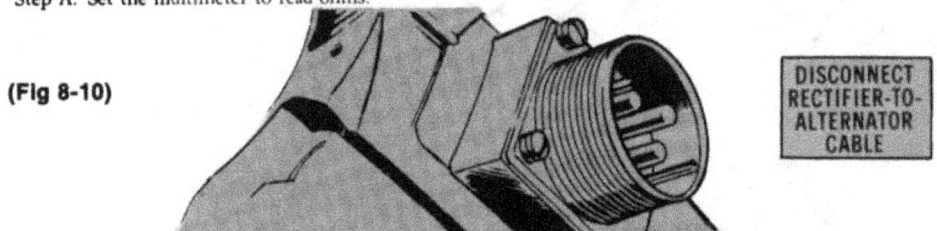

(Fig 8-10)

DISCONNECT RECTIFIER-TO-ALTERNATOR CABLE

Step B. Disconnect the rectifier-to-alternator cable at the alternator, (rectifier end already disconnected).

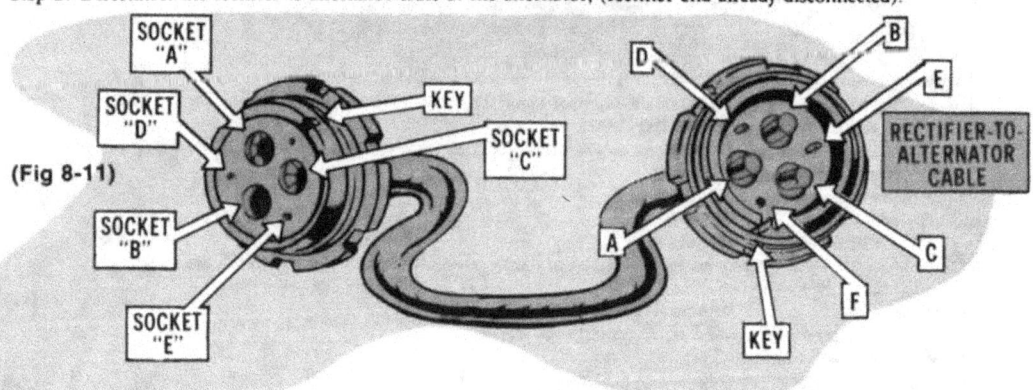

(Fig 8-11)

SOCKET "A" SOCKET "D" KEY SOCKET "C" SOCKET "B" SOCKET "E"

D B E RECTIFIER-TO-ALTERNATOR CABLE A C F KEY

Step C. Touch the black probe to pin A and the red probe to socket A. Touch the black probe to pin B and the red probe to socket B; touch the black probe to pin C, red probe to socket C; then black to pin D, red probe to socket D, then touch black probe to pin E, red probe to socket E.
   1. If you read 0 ohms each time, do Step D.
   2. If the needle does not read 0 ohms, repair or replace the rectifier-to-alternator cable. Repeat Check 5, page 9.

Step D. Touch the black probe to the cable connector shell and touch the red probe respectively to sockets A, B, C, D and E. (See Fig 8-11) Then, touch the black probe to ground and the red probe to Sockets A, B, C, D and E. (See Fig 8-11).
   1. If the needle does not move, replace the alternator. Repeat Check 5, on page 9.
   2. If the needle moves, repair or replace the rectifier-to-alternator cable. Repeat Check 5, page 9.

Test 6- Do the following steps to test the bulkhead feed thru-to-rectifier cable and the rectifier:

(Fig 8-12) DISCONNECT BULKHEAD-TO-RECTIFIER CABLE

Step A. Disconnect the bulkhead-to-rectifier cable at the bulkhead.
Step B. Touch the black probe to cable pin D and touch the red probe to pin C.

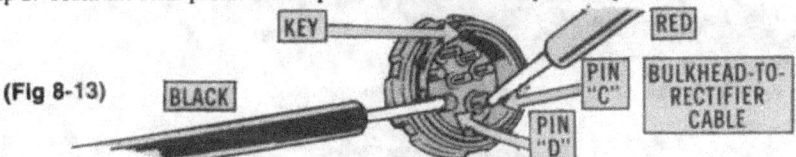

(Fig 8-13) KEY BLACK RED PIN "C" PIN "D" BULKHEAD-TO-RECTIFIER CABLE

   1. If the meter reads 55-75 ohms, repair or replace the bulkhead to regulator cable. Repeat Check 5, page 9.
   2. If the meter does not read 55-75 ohms, do Test 7.

Step C. Touch the black probe to large pin C and the red probe to large pin D. (See Fig 8-13)
   1. If the meter does not move, repair or replace the regulator to bulkhead cable. Repeat Check 5, page 9.
   2. If the needle moves, do Test 7.

Step D. Touch the black probe to medium pin B and touch the red probe to medium pin E. (See Fig 8-13).
   1. If the meter reads 1-4 ohms, repair or replace the regulator to bulkhead cable. Repeat Check 5, page 9.
   2. If the meter does not read 1-4 ohms, do Test 7.

Step E. Touch the black probe to medium pin B and touch the red probe to large pin C and then to large pin D. (see Fig 8-13).
   1. If the needle moves, do Test 7.
   2. If the needle does not move, repair or replace the regulator to bulkhead cable. Repeat Check 5, page 9.

Step F. Touch the black probe to ground, and touch the red probe to pins B, C, D and E.
   1. If the needle moves, do Test 7.
   2. If the needle does not move, do Step G.

Step G. Touch the black probe to the connector shell, and touch the red probe to pins B, C, D and E.
   1. If the needle moves the wires in the connector shell are shorted. Repair or replace the wires.
   2. If the needle does not move, repair or replace the regulator-to-bulkhead cable.

Test 7- Do the following steps to test the bulkhead-to-rectifier cable for shorts and continuity:
Step A. Disconnect the bulkhead-to-rectifier cable at the rectifier. Bulkhead end already disconnected.

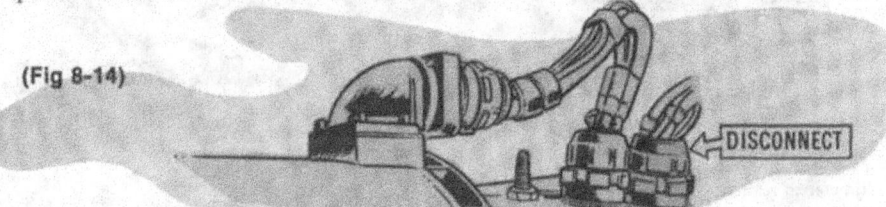

(Fig 8-14) DISCONNECT

Step B. Touch the black probe to socket B and the red probe to pin B. Touch the black probe to socket C, the red probe to pin C; touch the black probe to socket D, the red probe to pin D; then black probe to socket E, red probe to pin E.

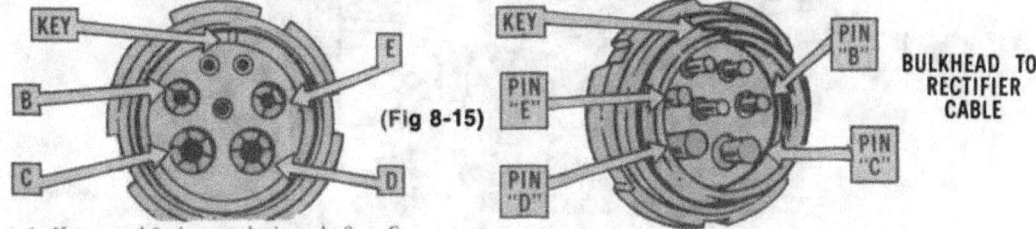

KEY E B C D **(Fig 8-15)** KEY PIN "E" PIN "D" PIN "B" PIN "C" BULKHEAD TO RECTIFIER CABLE

   1. If you read 0 ohms each time, do Step C
   2. If the needle does not read 0 ohms, repair or replace the bulkhead to rectifier cable. Repeat Check 5, page 9.

37

Step C. Touch the black probe to the connector shell and touch the red probe respectively to pins B, C, D, and E. (See Fig 8-15).
  1. If the needle does not move, replace the rectifier. Repeat Check 5, page 9.
  2. If the needle moves, repair or replace the bulkhead-to-rectifier cable.

Test 8- With the engine off do the following steps to test the generator field switch:

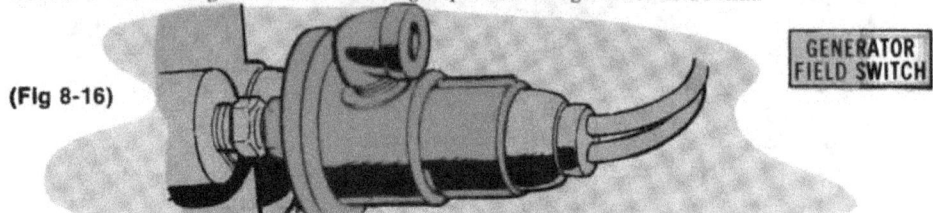

(Fig 8-16)

Step A. Remove the connector from the generator field switch. Using a short jumper lead, short the contact of A and B in the field switch cable plug.

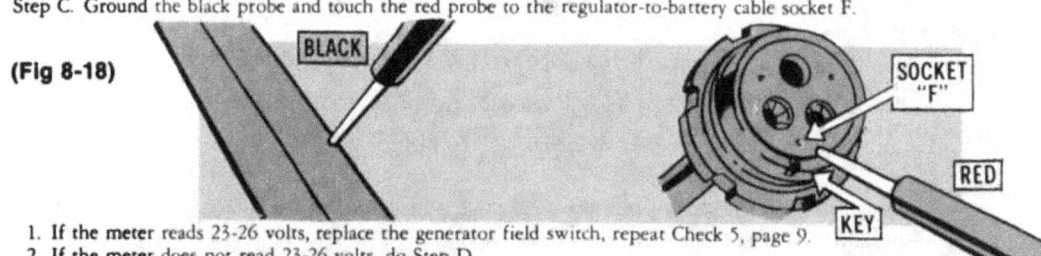

(Fig 8-17)

Step B. Set multimeter to read battery volts.
Step C. Ground the black probe and touch the red probe to the regulator-to-battery cable socket F.

(Fig 8-18)

  1. If the meter reads 23-26 volts, replace the generator field switch, repeat Check 5, page 9.
  2. If the meter does not read 23-26 volts, do Step D.
Step D. At the generator field switch ground the black probe to the vehicle and touch the red probe to cable socket A.

(Fig 8-19)

  1. If you read 0 volts, repair or replace cable 1A. Repeat Check 5, page 9.
  2. If you read 23-26 volts, do Step E.
Step E. Set multimeter to read ohms.
Step F. Start engine and run at 1000-1200 RPM.
Step G. Touch the black probe to pin A and the red probe to pin B of the generator field switch.

(Fig 8-20)

  1. If you read 0 ohms, repair or replace wire #1B between the field switch and the regulator. Repeat Check 5, page 9.
  2. If you do not read 0 ohms, replace the generator field switch. Repeat Check 5, page 9.

Chapter 9

100 AMP, EXTERNALLY RECTIFIED, MOLDED TYPE
(TRACK)

NOTE: *Do the Battery Voltage Checks on pages 8 and 9 before doing Test 1.*

Test 1- Do the following steps to test the rectifier-to-alternator cable and the alternator:

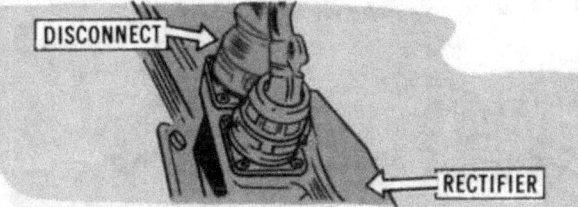

(Fig 9-1)

Step A. Disconnect the rectifier-to-alternator cable at the rectifier.
Step B. Set meter to read ohms (RX1, X1). "Zero" the meter.
Step C. Touch the black probe to large cable pin A and touch the red probe to large cable pins B and then C.

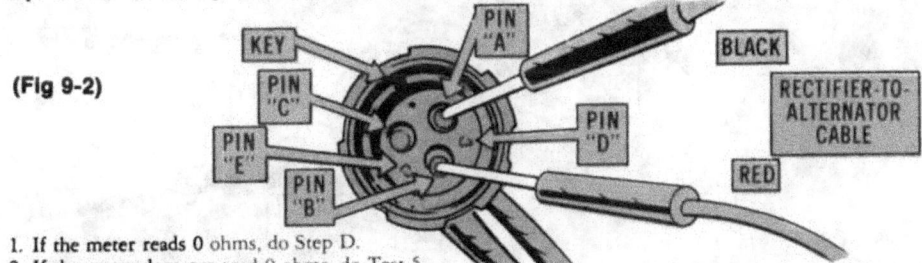

(Fig 9-2)

1. If the meter reads 0 ohms, do Step D.
2. If the meter does not read 0 ohms, do Test 5.
Step D. Touch the black probe to small cable pin D and touch the red probe to small cable pin E. (See Fig 9-2).
1. If you read 1-4 ohms, do Step E.
2. If you do not read 1-4 ohms, do Test 5.
Step E. Touch the black probe to large pin C and touch the red probe to small pin E. (See Fig 9-2).
1. If you read 0 ohms do Test 5.
2. If you do not read 0 ohms do Step F.
Step F. Touch the black probe to the cable connector shell and touch the red probe, in turn, to pins A, B, C, D and E.
1. If the needle does not move, do Step G.
2. If the needle moves, the wires in the connector shell are shorted. Repair or replace the wires.
Step G. Then touch the black probe to ground and the red probe to pins A, B, C, D and E. (See Fig 9-2).
1. If the needle moves, do Test 5.
2. If the needle does not move, the rectifier-to-alternator cable and alternator check okay. Reconnect the cable and do Test 2.

Test 2- Do the following steps to test the regulator-to-rectifier cable and the rectifier:
Step A. Disconnect the regulator-to-rectifier cable at the regulator.

(Fig 9-3)

Step B. Set meter to read ohms (RX1, X1). "Zero" the meter.
Step C. Touch the black probe to ground and touch the red probe to large pin C. (See Fig 9-3).
1. If the needle just barely moves or does not move at all, do Step D.
2. If the needle reads less than 1000 (1K ohms), do Test 6.
Step D. Now, reverse the probes. Touch the black probe to large pin C and touch the red probe to ground.
1. If the meter reads between 55-75 ohms, do Step E.
2. If the needle does not read 55-75 ohms, do Test 6.
Step E. Touch the black probe to medium pin B and touch the red probe to medium pin E.(See Fig 9-3).
1. If you read 1-4 ohms, do Step F.
2. If you do not read 1-4 ohms, do Test 6.

Step F. Touch the black probe to medium pin B and touch the red probe to large pin C, then to a good ground.
   1. If the needle does not move, do Step G.
   2. If the needle moves, do Test 6.
Step G. Touch the black probe to the connector shell, and touch the red probe to pins B, C, D and E.
   1. If the needle does not move, do Test 3.
   2. If the needle moves, the wires in the connector shell are shorted. Repair or replace the wires.

Test 3- Do the following steps to test the regulator-to-battery cable:

Step A. With engine off disconnect the regulator-to-battery cable at the regulator.

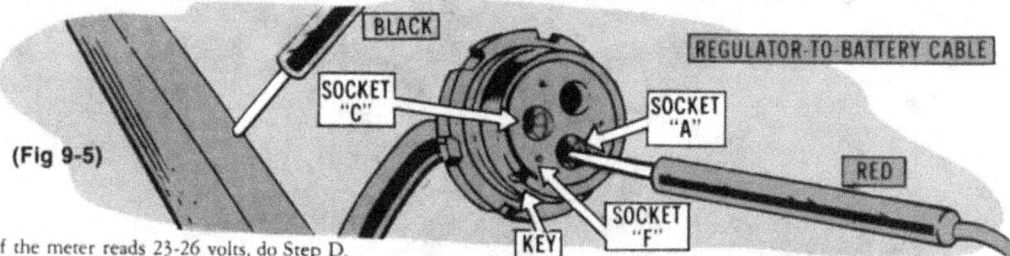

(Fig 9-4)

Step B. Set multimeter up to read battery volts. Turn on master switch.

Step C. Ground the black probe to the vehicle chassis and touch the red probe to large socket A of the regulator-to-battery cable.

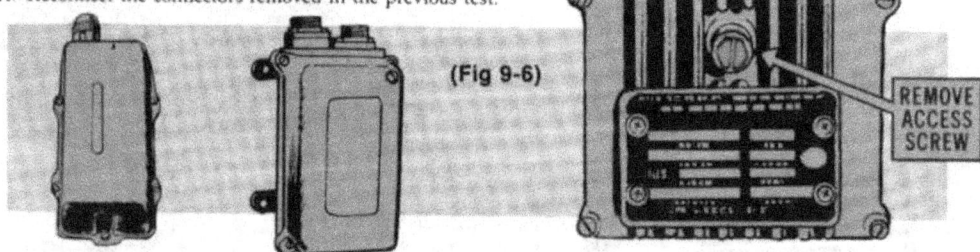

(Fig 9-5)

   1. If the meter reads 23-26 volts, do Step D.
   2. If the meter does not read 23-26 volts, repair or replace the cable from the regulator to the master relay. Repeat Check 5, page 9.

Step D. Set the meter to read ohms (RX1, X1). "Zero" the meter.

Step E. Touch the black probe to ground and touch the red probe to socket C of the cable. (See Fig 9-5).
   1. If the meter reads 0 ohms do Step F.
   2. If the meter does not read 0 ohms, repair or replace the cable from the regulator to its ground connection. Repeat Check 5, page 9.

Step F. Set meter to read battery volts.

Step G. Start the engine and run it at 1000 to 1200 RPM.

Step H. Ground the black probe and touch the red probe to socket F of the cable. (See Fig 9-5).
   1. If the meter reads 23-26 volts, turn engine off and do Test 4.
   2. If the meter does not read 23-26 volts, do Test 7.

Test 4- Do the following steps to adjust the regulator voltage output (if local SOP prohibits adjustment of the regulator at this level, replace the regulator and repeat Check 5, page 9.

Step A. Reconnect the connectors removed in the previous test.

(Fig 9-6)

REMOVE ACCESS SCREW

Step B. Remove the regulator adjustment access screw or the cover to expose the voltage adjustment screw in the regulator.

Step C. Start engine, turn on high beam lights and allow the engine to reach normal operating temperature.

Step D. After engine is warmed up, bring engine speed up to 1000-1200 RPM and hold it steady.

Step E. Set the multimeter up to read battery volts.

Step F. Connect the black probe to the vehicle battery or slave receptacle negative terminal and connect the red probe to the positive terminal.

40

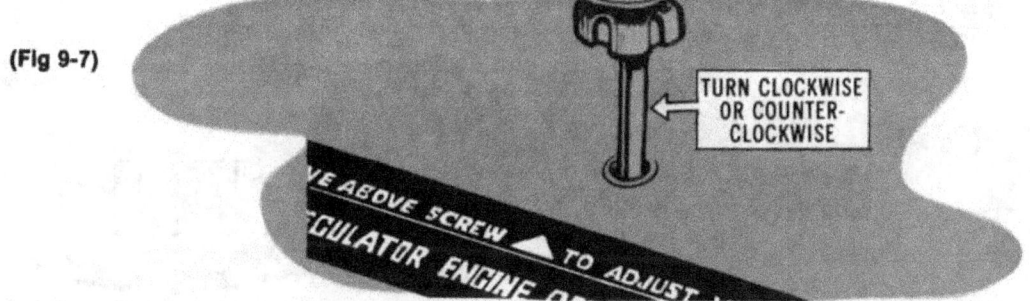

(Fig 9-7)

TURN CLOCKWISE
OR COUNTER-
CLOCKWISE

VE ABOVE SCREW ▲ TO ADJUST

GULATOR ENGINE C

Step G. Attempt to adjust the regulator voltage to 27-29 volts by turning the voltage adjust screw clockwise or counter-clockwise.

(Fig 9-8)

COVER REMOVED
(NOT NECESSARY
ON ALL MODELS)

VOLTAGE
ADJUSTMENT
SCREW

1. If the voltage will adjust and stabilize within the 27-29 volt range, the charging system now checks okay.
2. If the voltage will not adjust to 27-29 volt, replace the regulator and repeat Check 5, page 9.

Test 5- Do the following steps to test the rectifier-to-alternator cable for shorts and continuity:

Step A. Disconnect the rectifier-to-alternator cable at the alternator (rectifier end already disconnected).

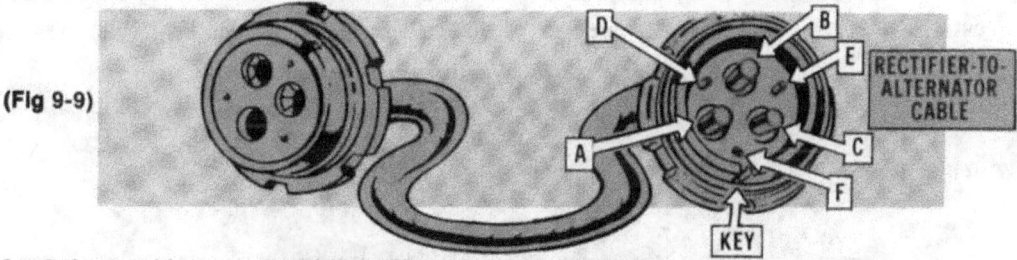

(Fig 9-9)

RECTIFIER-TO-
ALTERNATOR
CABLE

KEY

Step B. Leave multimeter set to read ohms (RX1, X1). "Zero" the meter.

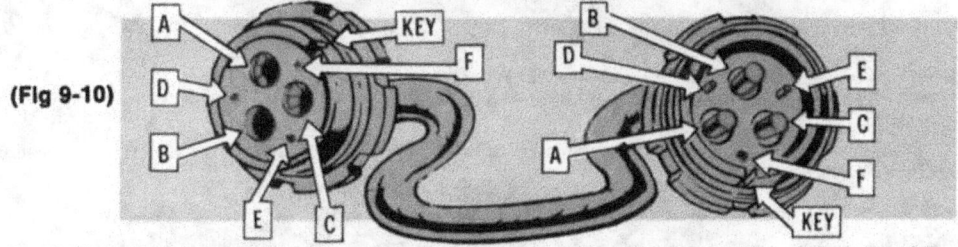

(Fig 9-10)

KEY

KEY

Step C. Touch the black probe to pins A, B, C, D and E and touch the red probe to sockets A, B, C, D and E.
1. If you read 0 ohms each time, do Step D.
2. If you do not read 0 ohms each time, repair or replace the cable. Repeat Check 5, page 9.

Step D. Touch the black probe to the connector shell and touch the red probe to sockets A, B, C, D and E. (See Fig 9-10).
1. If the needle does not move, the cable is okay. Replace the alternator. Repeat Check 5, page 9.
2. If the needle moves, the cable is shorted. Repair or replace the cable. Repeat Check 5, page 9.

Test 6 Do the following steps to test the regulator-to-rectifier cable for shorts and continuity:
Step A. Disconnect the regulator-to-rectifier cable at the rectifier (regulator end already loose).

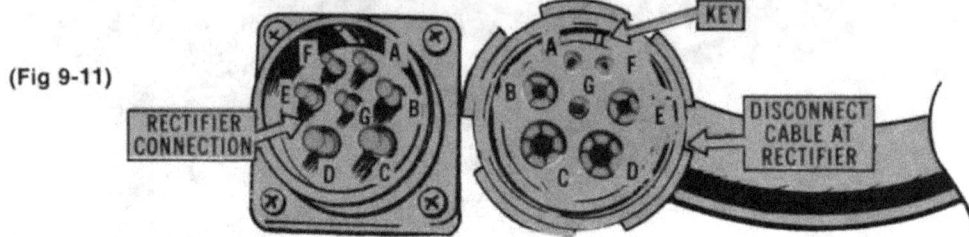

(Fig 9-11)

Step B. Set the meter to read ohms (RX1, X1). "Zero" the needle.
Step C. Touch the black probe to pins B, C and E (regulator end) and touch the red probe to sockets B, C and E (rectifier end).

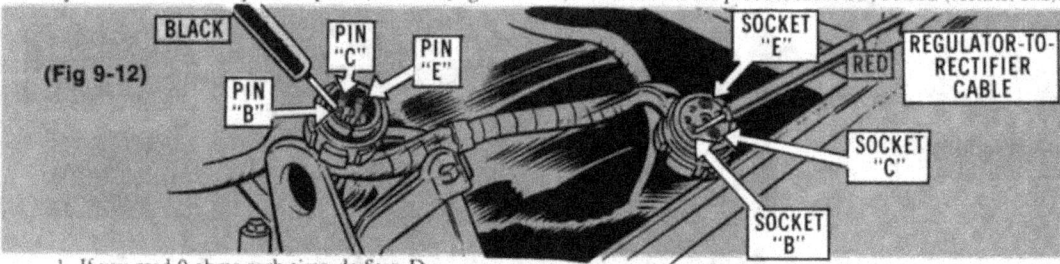

(Fig 9-12)

1. If you read 0 ohms each time, do Step D.
2. If you do not read 0 ohms each time, repair or replace the cable. Repeat Check 5, page 9.
Step D. Touch the black probe to the connector shell and touch the red probe to sockets B, C and E. (see Fig 9-11).
1. If the needle does not move, the cable is okay. Replace the rectifier. Repeat Check 5, page 9.
2. If the needle moves, repair or replace the cable. Repeat Check 5, page 9.

Test 7 Do the following steps to test the air cleaner dust exhauster blower motor relay, the relay switch and the wiring:
Step A. Observe the blower motors.

(Fig 9-13)

1. If the blowers are working, turn off the engine and do Step B.
2. If the blowers are not working, turn the engine off and do Step E.
Step B. Keep the meter set to read battery volts.
Step C. Disconnect the driver compartment-to-battery compartment bulkhead feed thru connector in the driver compartment (at driver's feet).

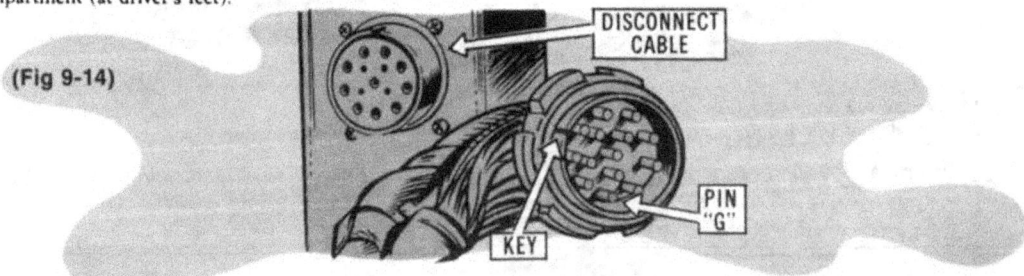

(Fig 9-14)

42

Step D. Ground the blakc probe and touch the red probe to cable pin G.

    1. If the meter reads 23-26 volts, repair or replace the cable between the bulkhead feed thru connector and the regulator.

    2. If the meter does not read 23-26 volts, repair or replace the cable between the bulkhead feed thru connector and the tie point behind the driver's panel (where wire 415 and 27B join together).

Step E. Disconnect the connector from the blower relay behind the driver's panel.

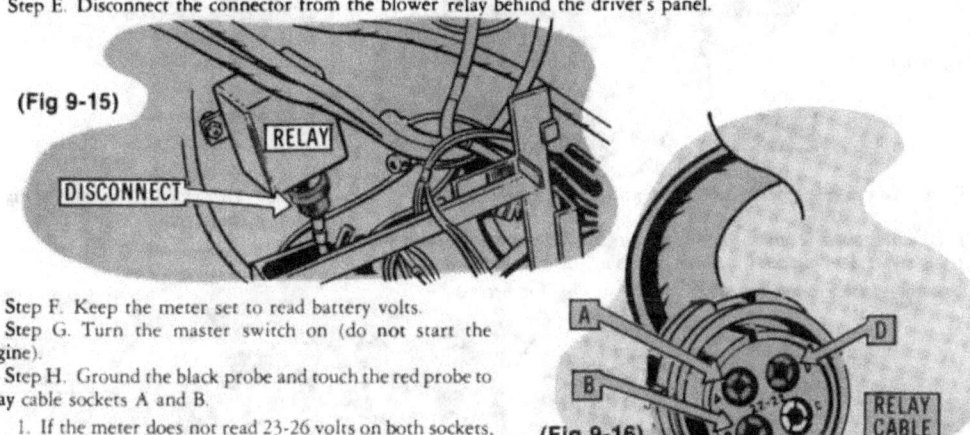

(Fig 9-15)

Step F. Keep the meter set to read battery volts.

Step G. Turn the master switch on (do not start the engine).

Step H. Ground the black probe and touch the red probe to relay cable sockets A and B.

    1. If the meter does not read 23-26 volts on both sockets, do Step I.

    2. If the meter reads 23-26 volts on each socket, do Step K.

    3. If the meter reads 23-26 volts on one socket but not on the other, repair or replace the wire to the defective socket.

Step I. Disconnect the #415 wire at the blower circuit breaker.

Step J. Ground the black probe and touch the red probe to the exposed terminal of the circuit breaker.

    1. If the meter reads 23-26 volts, repair or replace the #415 wire between the circuit breaker and the relay cable connector.

    2. If the meter does not read 23-26 volts, disconnect the #10 wire from the other side of the ciruit breaker. Ground the black and touch the red probe to the #10 wire.

      (a) If the meter reads 23-26 volts, replace the circuit breaker.

      (b) If the meter does not read 23-26 volts, repair or replace the #10 wire from the circuit breaker to the battery power.

Step K. Set the meter to read ohms (RX1, X1). "Zero" the meter.

Step L. Ground the black probe and touch the red probe to relay cable connector socket D.

    1. If the meter reads 0 ohms, replace the relay.

    2. If the meter does not read 0 ohms, do Step M.

Step M. Remove the driver compartment-to-engine compart inspection cover.

Step N. Start the engine.

Step O. Disconnect the relay switch cable plug from the switch.

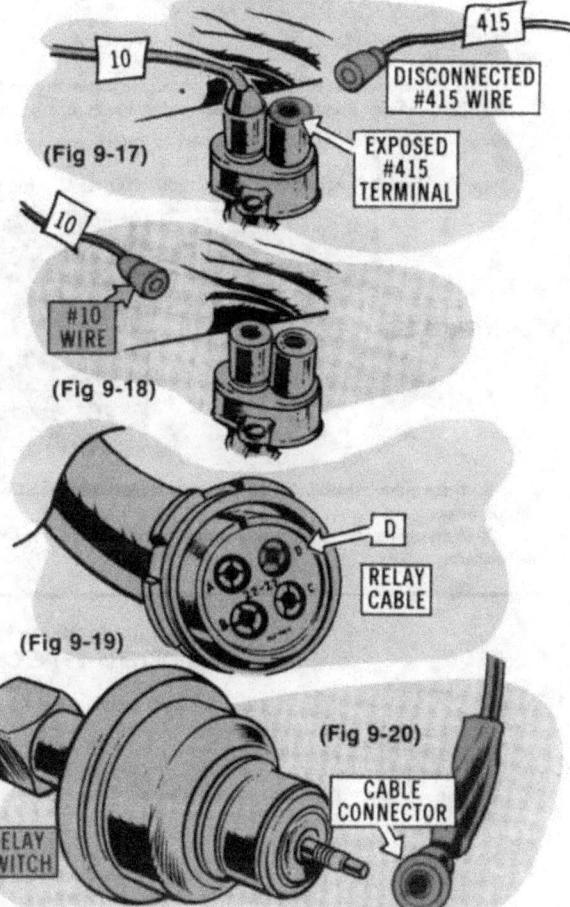

(Fig 9-16) RELAY CABLE

(Fig 9-17) 415 DISCONNECTED #415 WIRE EXPOSED #415 TERMINAL 10

(Fig 9-18) #10 WIRE 10

(Fig 9-19) RELAY CABLE D

(Fig 9-20) CABLE CONNECTOR RELAY SWITCH

43

Step O. Disconnect the relay switch cable plug from the switch.

(Fig 9-21)

Step P. With the engine at 1000-1200 RPM, ground the black probe and touch the red probe to the switch terminal. (See Fig 9-21).

   1. If the meter reads 0 ohms, the switch is okay. Reconnect the cable and do Step Q.
   2. If the meter does not read 0 ohms, the switch is bad. Replace it.

Step Q. Disconnect the engine disconnect at the transmission bracket.

(Fig 9-22)

Step R. With the engine running at 1000-1200 RPM, ground the black probe and touch the red probe to pin A on the bracket. (See Fig 9-22).

   1. If the meter reads 0 ohms, reconnect the cable and do Step S.
   2. If the meter does not read 0 ohms, repair or replace the #415B wire from the disconnect bracket to the relay switch.

Step S. Disconnect the driver-to-battery compartment cable at the bulkhead in the driver compartment at the driver's feet.

Step T. With the engine running at 1000-1200 RPM, ground the black probe and touch the red probe to socket A.

(Fig 9-23)

   1. If the meter reads 0 ohms, repair or replace the #415B wire that runs from the bulkhead cable connector to the blower motor relay.
   2. If the meter does not read 0 ohms, repair or replace the #415B wire that runs from the bulkhead feed thru to the engine disconnect.

**END OF 100 AMP, EXTERNALLY RECTIFIED,
MOLDED TYPE CHARGING TESTS
(TRACK)**

Chapter 10
M48's, M60's, M88's (300 AMP)

Test 1- Check to see if you have a solid state or a carbon pile regulator:

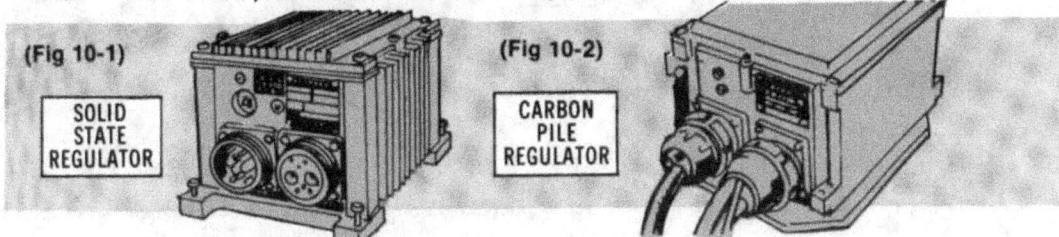

(Fig 10-1) SOLID STATE REGULATOR

(Fig 10-2) CARBON PILE REGULATOR

. 1. If you have a solid state regulator and:
    (a) In Check #5, page 9, the system was charging in the 27-29 volt range, but the batteries still get weak, do Test 8.
    (b) In Check #5 the system was not charging in the 27-29 volt range, do Test 2.
2. If you have a carbon pile regulator and:
    (a) In Check #5 the system was charging in the 27-29 volt range, but the batteries still get weak, do Test 8.
    (b) In Check #5 the system was not charging in the 27-29 volt range, do Test 3.

Test 2- Do the following steps to reset the solid state regulator overvoltage relay:

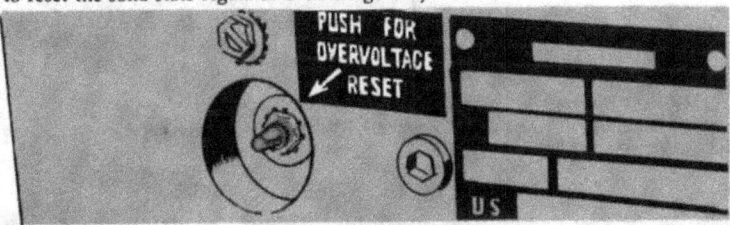

(Fig 10-3)

PUSH FOR OVERVOLTAGE RESET

US

Step A. Push the reset button.
Step B. Set the meter to read battery volts.
Step C. Start engine. Turn on high beam lights. Idle at 1000-1200 RPM. When the engine reaches operating temperature do Step D.
Step D. Measure the charging system output voltage at the batteries or at the slave receptacle.
    1. If the meter now reads 27-29 volts the charging system is okay.

NOTE: *If the overvoltage relay keeps tripping you must find the source of the overvoltage. First replace the regulator. If this doesn't solve the problem, do Test 3.*

    2. If the meter still does not read 27-29 volts, do Test 3.

Test 3- Do the following steps to test the regulator-to-battery cable for voltage.

(Fig 10-4)

REMOVE REGULATOR-TO-BATTERY CABLE

REGULATOR

Step A. Remove the regulator-to-battery cable at the regulator.
Step B. Keep the meter set for battery volts.
Step C. Turn the master switch on.
Step D. Touch the black probe to a good ground. Touch the red probe to Socket A of the cable.

Step D. Touch the black probe to a good ground. Touch the red probe to Socket A of the cable.

(Fig 10-5)

VEHICLE FRAME

BLACK

KEY

SOCKET "A"

RED

1. If the meter reads 23-26 volts, turn the switch off and do Test 4.
2. If the meter reads 0 volts, turn the master switch off.

(a) If the vehicle you are testing is an M60 tank, repair or replace the cable between the regulator and the master relay. Repeat Check 5, page 9.

(b) If the vehicle you are testing is an M88, do Step E.

Step E. Make sure that the master switch is off. Remove the battery cable from the "Little Joe."

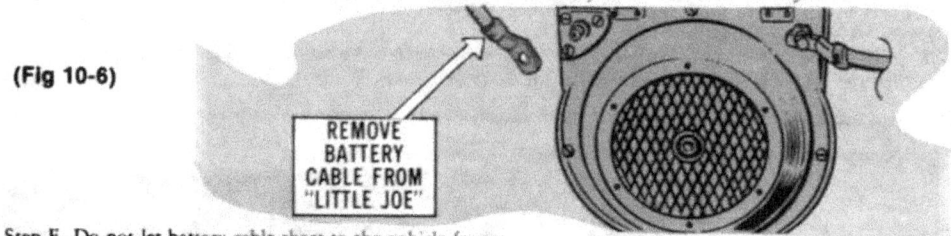

(Fig 10-6)

REMOVE BATTERY CABLE FROM "LITTLE JOE"

Step F. Do not let battery cable short to the vehicle frame.
Step G. Turn the master switch on.
Step H. Touch the black probe to ground and touch the red probe to the cable and to the master relay. Read the meter. Turn the master switch off.

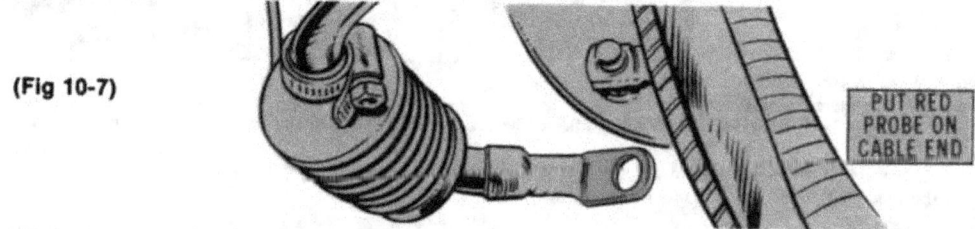

(Fig 10-7)

PUT RED PROBE ON CABLE END

1. If the meter reads 0 volts, repair or replace the cable between "Little Joe" and the master relay. Repeat Check 5, page 9.
2. If the meter reads 23-26 volts, do Step I.
Step I. Set the multimeter up to read ohms.

(Fig 10-8)

GROUND

BATTERY TERMINAL

Step J. Touch the black probe to ground and touch the red probe to the battery terminal of "Little Joe." Read the meter.
1. If the needle does not move, repair or replace the cable between the "Little Joe" battery terminal and the regulator. Repeat Check 5, page 9.
2. If the needle moves or reads 0, replace the "Little Joe." Repeat Check 5, page 9.

Public Domain, Google-digitized / http://www.hathitrust.org/access_use#pd-google

Test 4 Do the following steps to test the generator and the regulator-to-generator cable for ohms at the regulator:

(Fig 10-9)

Step A. With the engine off and master switch off, disconnect the regulator-to-generator cable at the regulator.

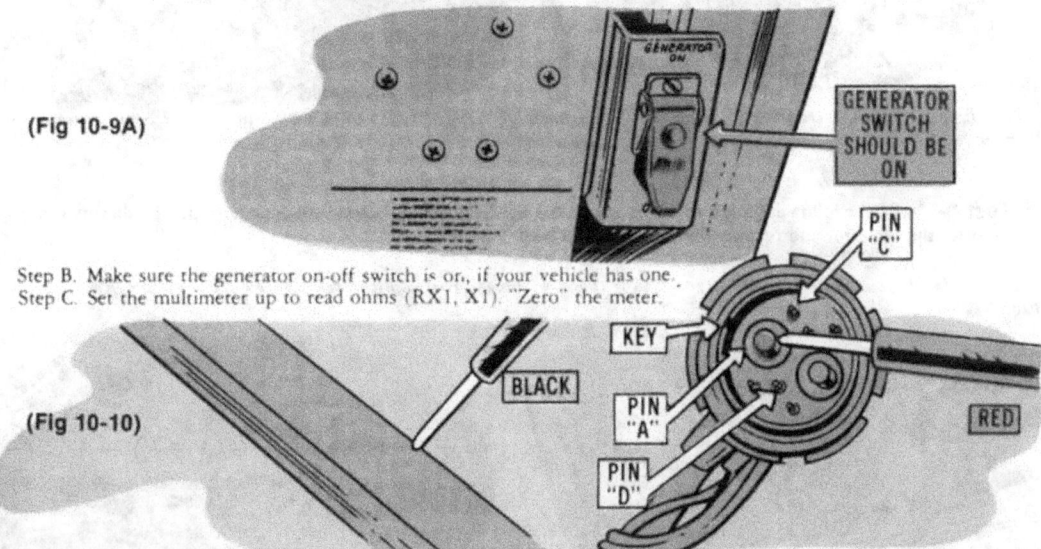

(Fig 10-9A)

Step B. Make sure the generator on-off switch is on, if your vehicle has one.
Step C. Set the multimeter up to read ohms (RX1, X1). "Zero" the meter.

(Fig 10-10)

Step D. Touch the black probe to a good ground. Touch the red probe to cable pin A, then to pin C, and pin D. Read the meter each time.
    1. If the meter reads 0 to 4 ohms on each pin, do Test 5.
    2. If the meter reads greater than 4 ohms on any of the pins and:
    (a) You are working on M60, do Test 9.
    (b) You are working on a M88, do Test 10.

Test 5 Do the following steps to test the generator for residual voltage:
Step A. Set multimeter to read 10 volts DC.
Step B. Start the engine and run at 1,000-1,200 RPM.
Step C. Ground the black probe to the vehicle frame and touch the red probe to pin A of the cable that goes from the regulator to the generator.

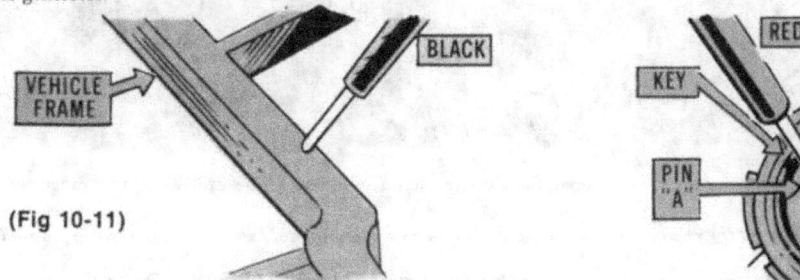

(Fig 10-11)

    1. If the meter reads between 1 and 4 volts, the generator is okay. Do Test 6 to see if adjusting the regulator remote rheostat will bring the volts up to the 27-29 volt range.

47

2. If the meter reads less than 1 volt or the needle deflects to the left of 0 volts (tries to go negative), do Steps D and E to polarize the generator.

3. If the meter reads more than 4 volts, replace the generator.

Step D. Polarizing the generator: Shut off the engine, (master switch on and generator on-off switch on for vehicles with a generator on-off switch, otherwise leave the master switch off). Use a heavy insulated jumper wire to jump from the middle battery cable (either terminal) to pin D on the regulator-to-generator cable. Repeat Steps B and C.

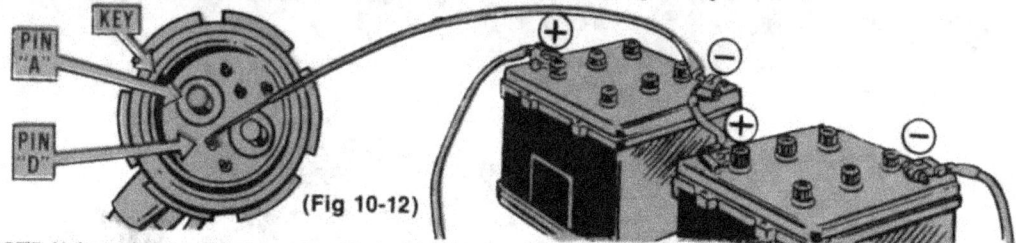

(Fig 10-12)

NOTE: If the reading is still bad (less than 1 volt or deflects left) repeat Step D. Do not attempt to polarize the generator more than twice. If the reading is bad after the second attempt, replace the generator. Repeat Check 5, page 9.

Test 6- Do the following steps to test or adjust the regulator voltage output if local SOP prohibits adjustment at your level replace the regulator and remote rheostat. Repeat Check 5 page 9.

NOTE: (a) If you have the solid state regulator there are no external adjustments. Replace the regulator. Repeat Check 5, page 9.

(b) If you have the carbon pile regulator, do Step A.

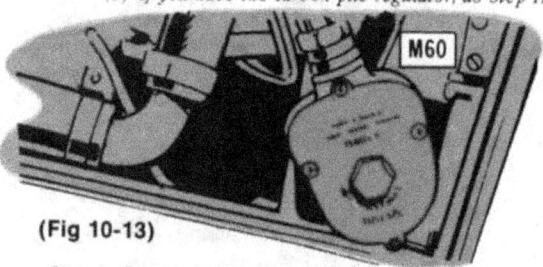

(Fig 10-13)

(Fig 10-14)

Step A. Reconnect all cable assemblies.
Step B. Set multimeter up to read battery volts.
Step C. Start engine. Turn on high beam lights. Idle at 1,000-1,200 RPM. When engine reaches normal operating temperature, do Step D.
Step D. Ground the black probe, and touch the red probe to the positive terminal of the batteries or the slave receptacle.
1. If the needle gives a steady reading in the 27-29 volt range the charging system checks out okay.
2. If the needle is not steady, adjust the regulator output as in Step E.

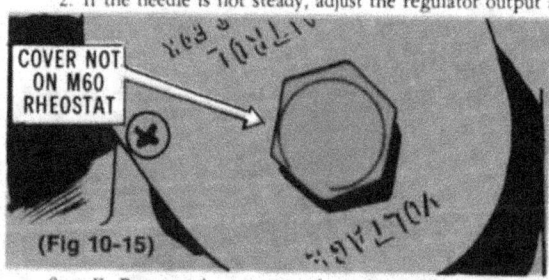

(Fig 10-15)

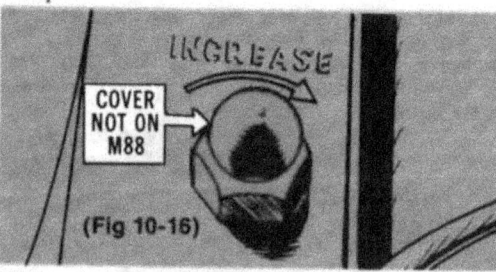

(Fig 10-16)

Step E. Remove the cover nut from the regulator remote rheostat.
Step F. Turn the screw with your fingers or a small flat tip screwdriver until you get a steady 27-29 volt range reading.

NOTE: During hot weather, set the reading toward the low end of the range. During cold weather set the reading toward the high end of the range.
1. *If you cannot get a steady reading in the 27-29 volt range, turn the engine and lights off. Do Test 7.*
2. *If you get a steady reading in the 27-29 volt range, the charging system now checks out okay.*

48

Test 7- Do the following steps to test the regulator remote rheostat:

NOTE: This test does not apply to vehicles with the solid state regulator.

Step A. With engine off and master switch off, remove the regulator-to-battery connector at the regulator.

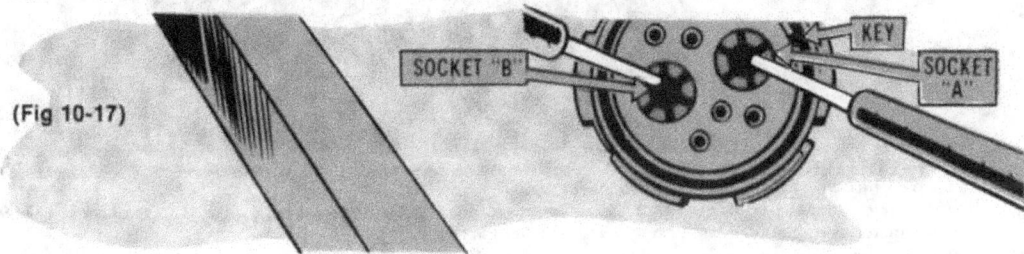

(Fig 10-17)

Step B. Set the multimeter up to read ohms (RX1, X1). "Zero" the meter.

Step C. Touch the black probe to cable socket B and the red probe to socket D. (See Fig 10-17).

Step D. Slowly turn the adjust-screw on the remote rheostat with a small flat tip screwdriver or your fingers. (See Fig 10-16). Turn it fully clockwise and fully counter clockwise while you watch the meter needle.

    1. If the needle swings from 0 to 100 ± 10 ohms the rheostat checks good. Replace the regulator. Repeat Test 6, page 48.

    2. If the needle does not swing 0 to 100 ± 10 ohms, repair the rheostat leads or replace the rheostat. Repeat Test 6, page 48.

Test 8- Reverse Current Relay Voltage can be checked by doing the following steps:

Step A. Disconnect the regulator-to-generator cable at the regulator.

Step B. Set the meter to read battery volts.

Step C. Turn on master switch.

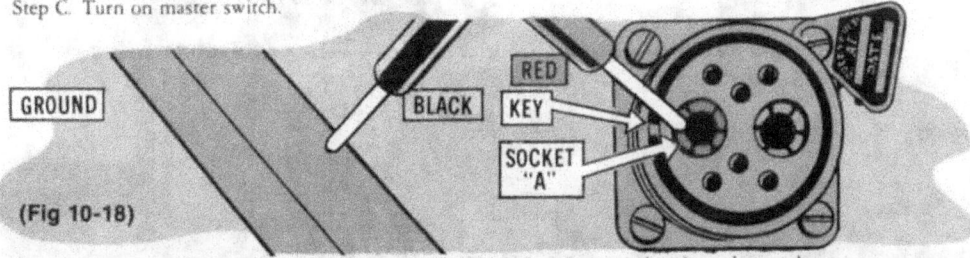

(Fig 10-18)

Step D. Touch the black probe to a good ground and the red probe to socket A on the regulator.

    1. If you read 0 volts:

       (a) You possibly have operator problems that are causing your batteries to be weak. They might be operating commo or weapons systems without running the engine or are leaving switches on when the vehicle is not being used.

       (b) The batteries may be defective. Do Checks #3 and 4, page 9, to be sure.

    2. If you read any volts, replace the regulator. Repeat Check 5, page 9.

Test 9- Do the following steps to test the generator and the bulkhead-to-generator cable at the bulkhead feed thru:

Step A. Disconnect the bulkhead-to-generator cables at the bulkhead feed thru.

(Fig 10-19)

Step B. Set the multimeter up to read ohms (RX1, X1). "Zero" the meter.

(Fig 10-20) RED KEY BLACK M60 BULKHEAD FEEDTHRUS

SOCKET "C"

PIN "B"

SOCKET "L"

KEY

Step C. Touch the black probe to a good ground. Touch the red probe to pin B, in the bottom connector, then to socket C and socket L in the other connector. Read the meter each time.

1. If the meter reads 0-4 ohms each time, repair or replace the **regulator-to-bulkhead** cable. Repeat Check 5, page 9.

2. If the meter reads greater than 4 ohms on any of the pins do Test 10.

Test 10- Do the following steps to test the generator and the **engine disconnect-to-generator** cable for ohms:

Step A. Remove disconnect cables at the disconnect. (See Fig 10-21 for M60 and Fig 10-22 for M88).

Step B. For the M60, touch the black probe to a good ground. Touch the red probe to the big socket. Then, touch it to pin C and pin L on the other connector. Read the meter each time. See 1 and 2, Step C.

(Fig 10-21) KEY BIG SOCKET M60 DISCONNECTS

PIN "L"

PIN "C"

Step C. On the M88, touch the black probe to a ground. Touch the red probe to socket A and B and then to the big socket. **Read** the meter each time.

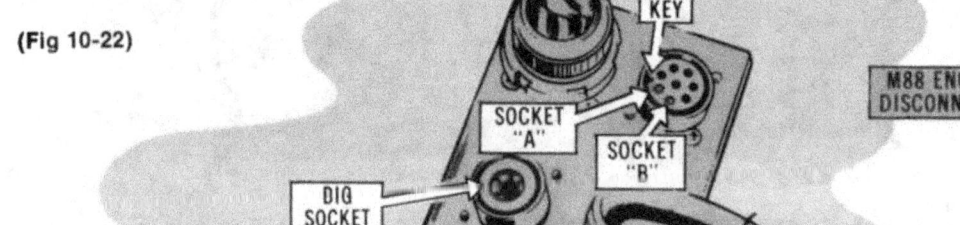

(Fig 10-22) KEY M88 ENGINE DISCONNECTS

SOCKET "A"

SOCKET "B"

BIG SOCKET

1. If the meter reads 0-4 ohms each time on the M60 replace the bulkhead-to-disconnect cable. On the M88 replace **the** disconnect-to-regulator cable. Repeat Check 5, page 9.

2. If the meter reads greater than 4 ohms on any of the pins, either the rest of the cable assembly in the engine compartment or the generator is bad. Do Test 11.

.Test 11- Pull the pack and do the following tests to check the engine disconnect-to-generator cable:

NOTE: Be sure cable connections on the generator are clean and tight. If they are, do Step A of this test. If they are not, clean and tighten as necessary and repeat Test 10 on page 50 with the pack out.

Step A. Remove the disconnect-to-generator cable wires at the generator.

(Fig 10-23) TERMINALS

Wire #1 goes to terminal A of the generator.
Wire #2 goes to terminal B of the generator.
Wire #478 goes to terminal D of the generator.
Wire #GND goes to terminal E of the generator.

Step B. Keep the meter set for ohms.

50

Step C. To check for continuity and shorts on the disconnect-to-generator cables, do the following: For the M60, do Step D. For the M88, do Step F.

Step D. To check for continuity, touch the black probe to pins C, L and the big socket. Touch the red probe, in turn, to wires #1, #478 and #2.

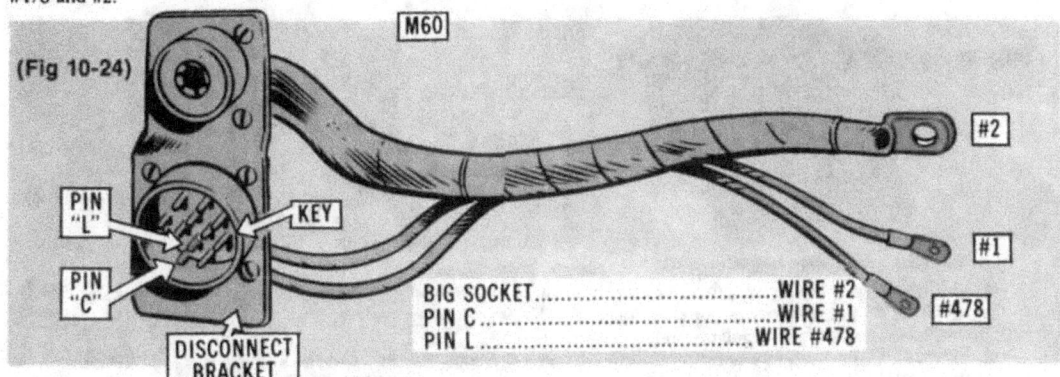

1. If the needle reads zero ohms, do Step E.
2. If the needle does not read zero ohms, repair or replace the defective wires.

Step E. To check for shorts, touch the black probe to wire #1, and touch the red probe to wires #478 and #2. Then, touch the black probe to wire #478 and touch the red probe to wire #2.
1. If the needle does not move, the cables and connections are okay. The generator is defective. Replace the generator and repeat Check 5, page 9.
2. If the needle moves, repair or replace the defective wires. Repeat Check 5, page 9.

Step F. To check for continuity, touch the black probe to sockets A and B and the big socket. In turn, touch the red probe to wires #1, #478 and #2.

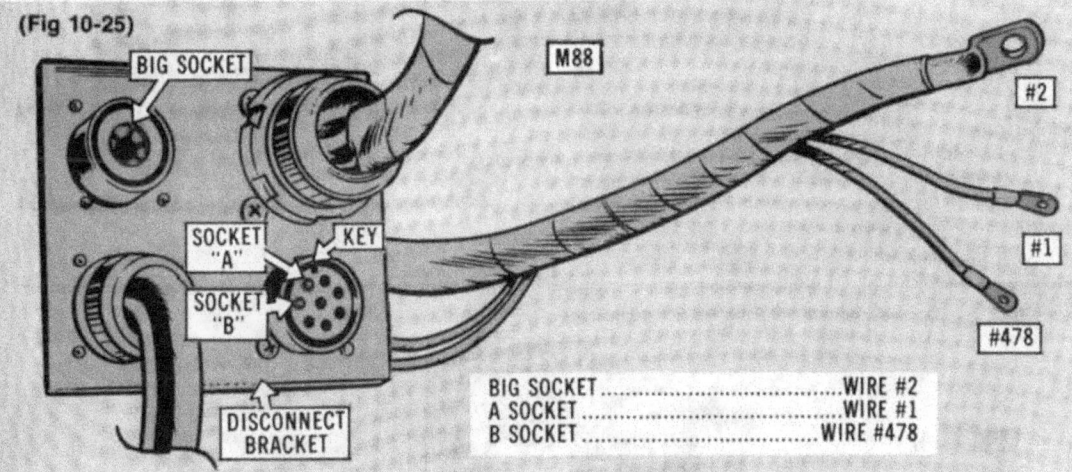

1. If the needle reads zero ohms, do Step G.
2. If the needle does not read zero ohms, repair or replace the defective wires.

Step G. To check for shorts, touch the black probe to wire #1 and touch the red probe to wires #478 and #2. Then, touch the black probe to wire #478 and touch the red probe to wire #2.
1. If the needle does not move, the cables and connections are okay. The generator is defective. Replace the generator and repeat Check 5 on page 9.
2. If the needle moves, repair or replace the defective wires. Repeat Check 5 on page 9.

Chapter 11

M107, M110, M578 (300 AMP)

Test 1- Check to see if you have a solid state or a carbon pile regulator:

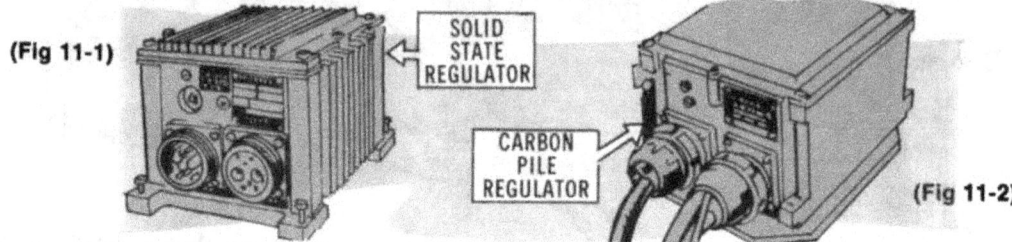

(Fig 11-1)

(Fig 11-2)

1. If you have a solid state regulator and:
   (a) In Check #5 page 9, the system was charging in the 27-29 volt range, but the batteries still get weak, do Test 8.
   (b) In Check #5 the system was not charging in the 27-29 volt range, do Test 2.
2. If you have a carbon pile regulator and:
   (a) In Check #5 the system was charging in the 27-29 volt range, but the batteries still get weak, do Test 8.
   (b) In Check #5 the system was not charging in the 27-29 volt range, do Test 3.

Test 2- Do the following steps to reset the solid state regulator over voltage relay:

(Fig 11-3)

Step A. Push the reset button.
Step B. Set the meter to read battery volts.
Step C. Start the engine and run it at 1000-1200 RPM with the high beam lights on until the engine reaches operating temperature.
Step D. Measure the charging system output voltage at the batteries or at the slave receptacle.
   1 If the meter now reads 27-29 volts the charging system is okay.
   NOTE: If the over voltage relay keeps tripping you must find the source of the over voltage. First replace the regulator. If this doesn't solve the problem do Test 3.

   2. If the meter still does not read 27-29 volts, do Test 3.

Test 3- Do the following steps to test the regulator-to-battery cable for voltage:

(Fig 11-4)

Step A. Remove the regulator-to-battery cable at the regulator.
Step B. Keep the meter set for battery volts.
Step C. Turn the master switch on.
Step D. Touch the black probe to a good ground. Touch the red probe to socket G of the cable.

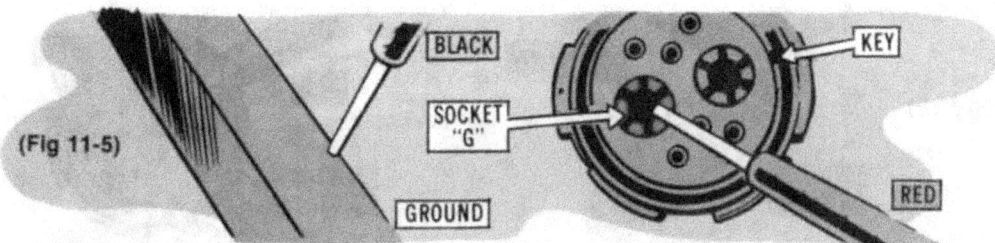

(Fig 11-5)

BLACK

KEY

SOCKET "G"

GROUND

RED

1. If the meter reads 23-26 volts, turn the master switch off. Then do Test 4.
2. If the meter does not read 23-26 volts. Repair or replace the cable between the regulator and the master relay.

Test 4- Do the following steps to test the generator and the regulator-to-generator cable for ohms at the regulator:

(Fig 11-6)

KEY

DISCONNECT CABLE

PIN "C"

PIN "G"

PIN "D"

Step A. With the engine off and master switch off, disconnect the regulator-to-generator cable at the regulator.
Step B. Make sure that the generator on-off switch is on . . . if you have one.

(Fig 11-7)

MUST BE ON!

Step C. Set the multimeter up to read ohms, (RX1, X1). "Zero" the meter.

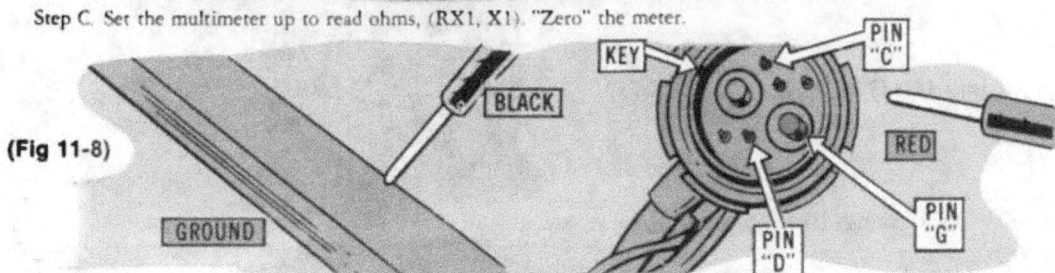

(Fig 11-8)

PIN "C"

KEY

BLACK

RED

PIN "G"

GROUND

PIN "D"

Step D. Touch the black probe to a good ground. Touch the red probe to cable pin C, then to pin D, and pin G. Read the meter each time.
1. If the meter reads 0 to 4 ohms on each pin, do Test 5.
2. If the meter reads greater than 4 ohms on any of the pins do Test 9.

Test 5- Do the following steps to test the generator for residual voltage:

Step A. Set multimeter to read 10 volts DC.
Step B. Start the engine and run at 1000-1200 RPM.
Step C. Ground the black probe to the vehicle frame, and touch the red probe to pin G of the cable that goes from the regulator to the generator.

(Fig 11-9)

1. If the meter reads between 1 and 4 volts, the generator is okay. Do Test 6 to see if adjusting the regulator remote rheostat will bring the volts up to the 27 to 29 volt range.

2. If the meter reads less than 1 volt or the needle deflects to the left of 0 volts (tries to go negative), do Steps D and E to polarize the generator.

3. If the meter reads more than 4 volts, replace the generator.

Step D. Polarizing the generator: Shut off the engine, master switch on and generator on-off switch on for vehicles with a generator on-off switch. Otherwise leave the master switch off. Use a heavy insulated jumper wire to jump from the middle battery cable (either terminal) to pin D on the regulator to generator cable. Repeat Steps B and C.

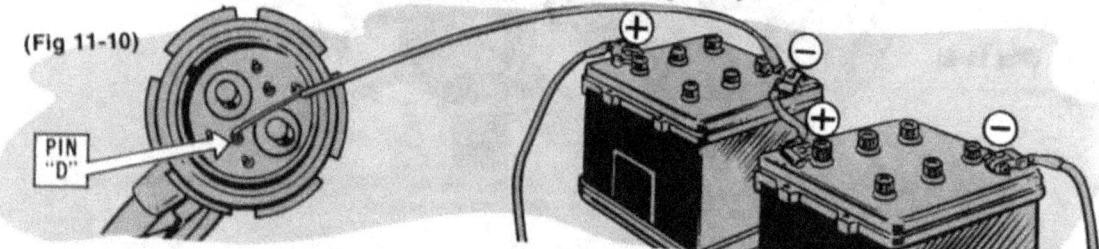

(Fig 11-10)

PIN "D"

NOTE: If the reading is still bad (less than .5 volt or deflects left) repeat Step D. Do not attempt to polarize the generator more than twice. If the reading is bad after the second attempt, replace the generator. Repeat Check 5, page

---

NOTE: If local SOP prohibits adjustment at your level, replace the regulator and remote rheostat. Repeat Check 5, page 9.

Test 6- Do the following steps to test or adjust the regulator voltage output:

NOTE: (a) If you have the solid state regulator there are no external adjustments. Replace the regulator.

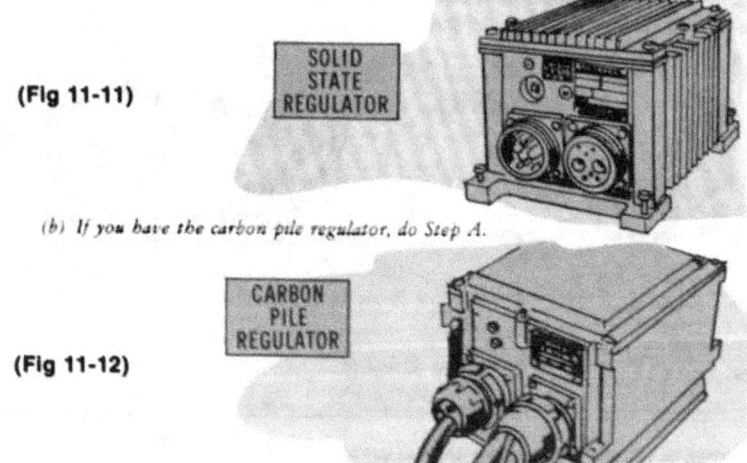

(Fig 11-11)

SOLID STATE REGULATOR

(b) If you have the carbon pile regulator, do Step A.

CARBON PILE REGULATOR

(Fig 11-12)

Step A. Reconnect all cable assemblies.
Step B. Set multimeter up to read battery volts.
Step C. Start engine. Turn on high beam lights. Idle at 1000-1200 RPM. When engine reaches normal operating temperature, do Step D.

54

Step D. Measure the charging system output voltage at the batteries or at the slave receptacle.
   1. If the needle gives a steady reading in the 27-29 volt range the charging system checks out okay.
   2. If the needle is not in the 27-29 volt range or is not steady, adjust the regulator output as in Step E.
Step E. Remove the cover nut from the regulator remote rehostat.

(Fig 11-13)

Step F. Turn the screw with your fingers or a small flat tip screwdriver until you get a steady 27-29 volt range reading.

   NOTE: *During hot weather, set the reading toward the low end of this range. During cold weather set the reading toward the high end of this range.*
   1. *If you cannot get a steady reading in the 27-29 volt range, turn the engine and lights off. Do Test 7.*
   2. *If you get a steady reading in the 27-29 volt range, the charging system now checks out okay.*

Test 7- Do the following steps to test the regulator remote rheostat:

Step A. With engine off and master switch off, remove the regulator-to-battery connector at the regulator.

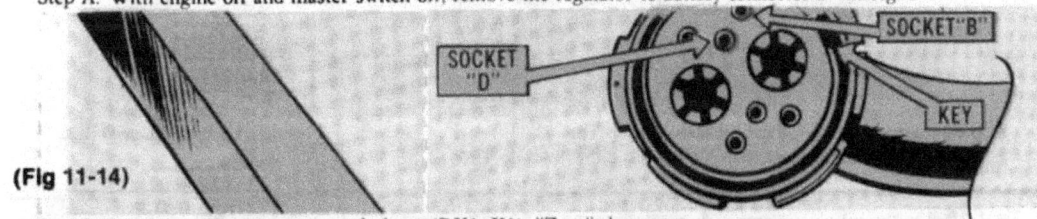

(Fig 11-14)

Step B. Set the multimeter up to read ohms, (RX1, X1). "Zero" the meter.
Step C. Touch the black probe to cable socket B and the red probe to socket D. (See Fig 11-14).
Step D. Slowly turn the adjust-screw on the remote rheostat with a small flat tip screwdriver or your fingers. Turn it fully clockwise and fully counter clockwise while you watch the meter needle.
   1. If the needle swings from 0 to 100 ± 10 ohms the rheostat checks good. Replace the regulator. Repeat Test 6, page 54.
   2. If the needle does not swing 0 to 100 ± 10 ohms, repair the rheostat leads or replace the rheostat. Repeat Test 6, page 54.

Test 8- Reverse Current Relay Voltage can be checked by doing the following steps:

Step A. Disconnect the regulator-to-generator cable.
Step B. Set meter to read battery volts.
Step C. Turn on master switch.
Step D. Touch the black probe to a good ground and the red probe to socket G on the regulator.

(Fig 11-15)

55

1. If you read 0 volts:
  (a) You possibly have operator problems that are causing your batteries to be weak. They might be operating commo or weapons systems without running the engine or are leaving switches on when the vehicle is not being used.
  (b) The batteries may be defective. Do Checks 3 and 4, page 9, to be sure.
2. If you read any volts, replace the regulator. Repeat Check 5, page 9.

Test 9- Do the following steps to test the generator and the #1 and #478 wires from the generator:

(Fig 11-16)

Step A. Disconnect the cable connerctor at the engine disconnect bracket located on the engine bulkhead.
Step B. Set the meter to read ohms (RX1, X1). "Zero" the meter.
Step C. Ground the black probe and touch the red probe to bulkhead disconnect connector pin A and then to pin B.

(Fig 11-17)

1. If the meter reads 0-4 ohms on each pin, do Test 10.
2. If the meter reads greater than 4 ohms on either pin, do Test 12.

Test 10- Do the following steps to test the bulkhead feedthru-to-disconnect bracket cable and large cable #2 from the bulkhead:
Step A. Reconnect the disconnect-to-bulkhead feedthru cable

(Fig 11-18)

Step B. Disconnect the bulkhead feedthru cable and large cable #2 from the bulkhead.

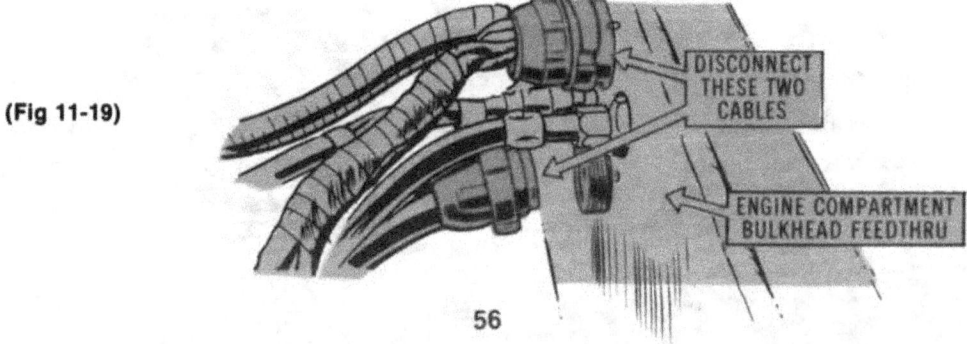

(Fig 11-19)

56

Step C. Keep the meter set to read ohms. "Zero" the meter.

Step D. Ground the black probe and touch the red probe to pin F and then to pin H. Then, touch the red probe to the large socket of cable #2 that goes to the generator. Read the meter each time.

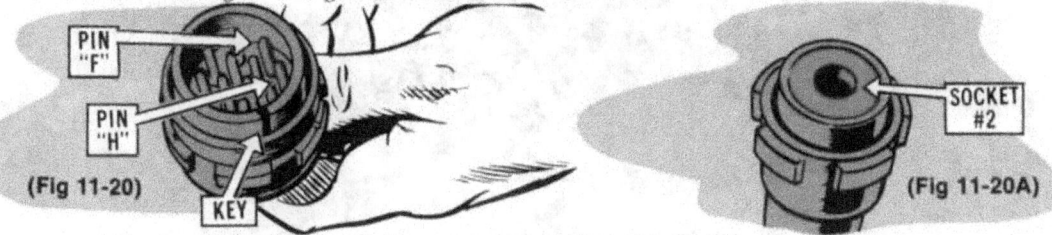

(Fig 11-20)

(Fig 11-20A)

1. If the meter reads 0-4 ohms each time, reconnect the cables and do Test 11.

2. If the meter reads more than 4 ohms on pins F and H, repair or replace the cable between the bulkhead feedthru and the disconnect bracket.

3. If the meter reads more than 4 ohms on large socket #2, do Test 12.

Test 11- Do the following steps to test the cables from the bulkhead feedthru (behind the driver compartment) to the engine compartment bulkhead feedthru:

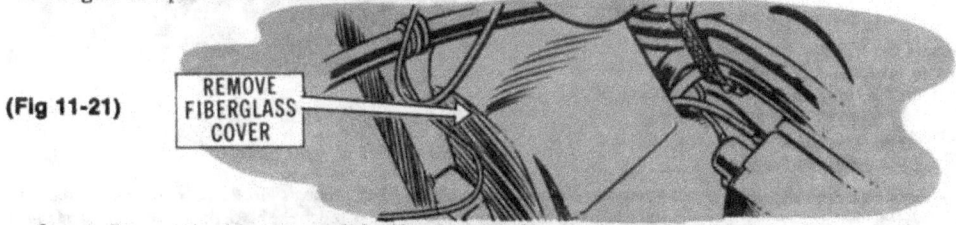

(Fig 11-21)

Step A. Remove the driver seat and the fiberglass covering.

Step B. Disconnect the bulkhead feedthru cables from behind the driver seat. (See Fig 11-22).

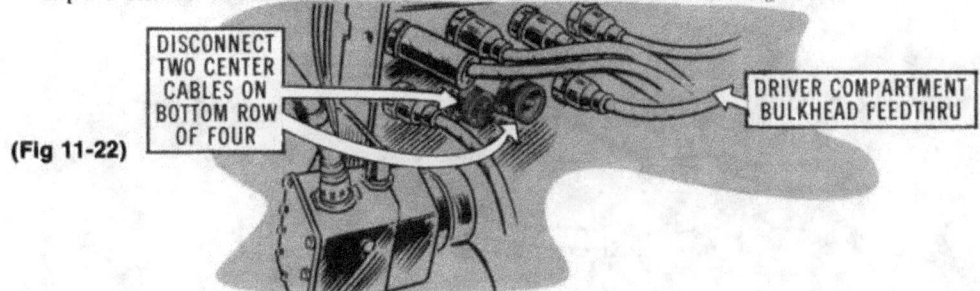

(Fig 11-22)

Step C. Set the meter to read ohms (RX1, X1). "Zero" the meter.

Step D. Ground the black probe and touch the red probe to cable pins A and then B. Then, touch the red probe to cable socket #2. Read the meter each time.

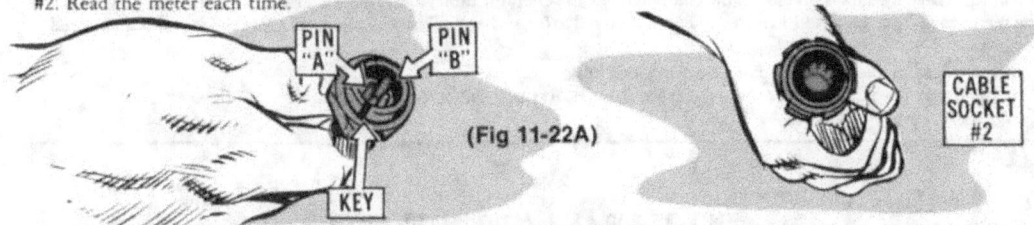

(Fig 11-22A)

1. If the meter reads 0-4 ohms on pins A and B and on socket #2, repair or replace wires #1 or #478 or cable #2 between the driver compartment bulkhead feedthru and the regulator.

2. If the meter reads greater than 4 ohms on pins A and B, repair or replace defective wire #1 or #478 between the driver compartment bulkhead to the engine bulkhead.

3. If the meter reads greater than 4 ohms on socket #2, repair or replace cable #2 between the driver's bulkhead and the engine bulkhead.

57

Test 12- Pull the pack and do the following steps to check the engine compartment -to-generator cables:

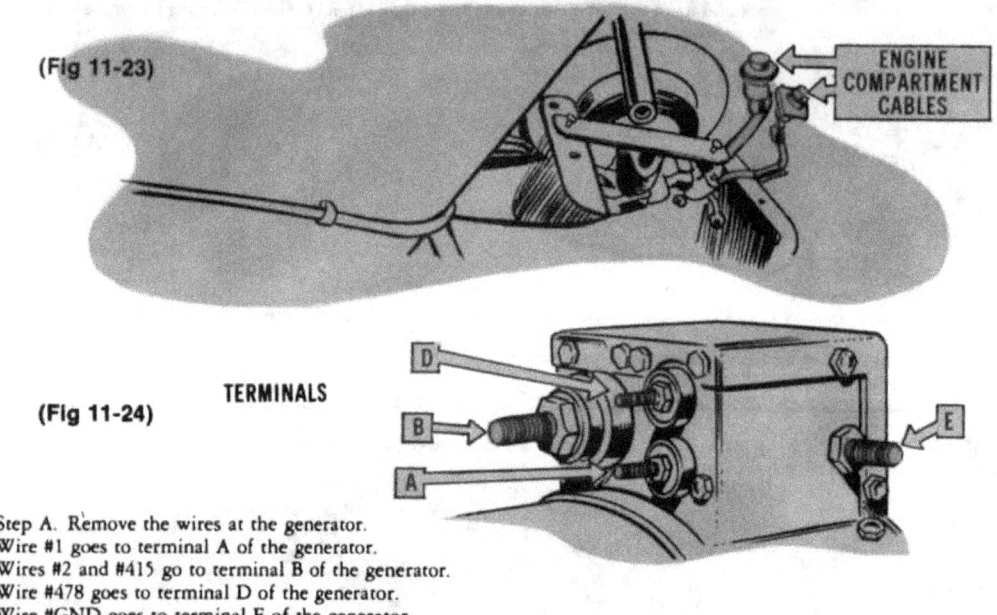

(Fig 11-23)

TERMINALS

(Fig 11-24)

Step A. Remove the wires at the generator.
Wire #1 goes to terminal A of the generator.
Wires #2 and #415 go to terminal B of the generator.
Wire #478 goes to terminal D of the generator.
Wire #GND goes to terminal E of the generator.
 Step B. Keep the meter set for ohms.

 Step C. To check for continuity on the engine compartment cables, touch the black probe to wire #1 and touch the red probe to pin B. Then, touch the black probe to wire #478 and touch the red probe to pin A. Then, touch the black probe to wire #2 and touch the red probe to socket #2. Note the meter reading each time.

(Fig 11-25)

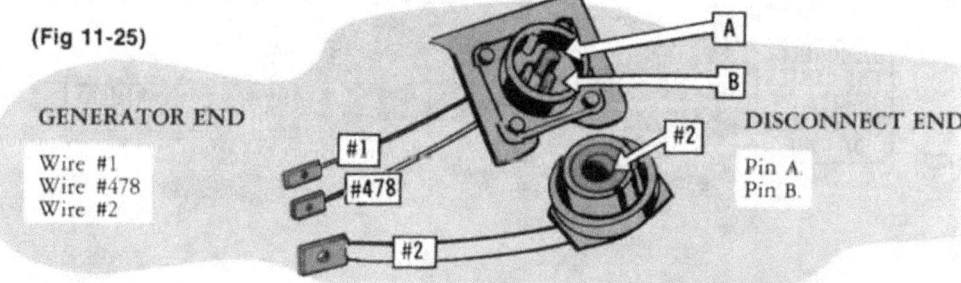

1. If the meter reads 0 ohms each time, do Step D to test for shorts.
2. If the meter does not read 0 ohms each time, repair and replace the cable.

 Step D. To check for shorts on the engine compartment to generator cable, touch the meter probes to the following pins and read the meter each time. Black to Pin A, red to Pin B, then red to big socket #2. Then, touch the black probe to Pin B and the red to big socket #2.

1. If the meter needle moves, repair or replace the cable.
2. If the meter needle does not move, the cable is good. Replace the defective generator.

**END OF 300 AMP, M107, M110, M578
CHARGING SYSTEM TESTS**

58

Chapter 12
M551 (300 AMP)

Test 1- Do the following steps to test the regulator-to-battery cable:

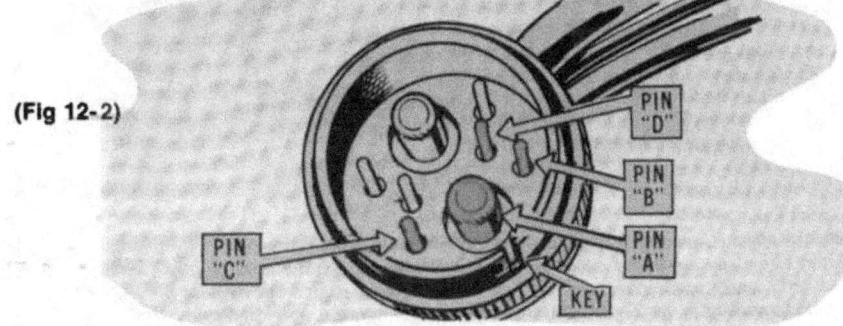

(Fig 12-1)

BLACK

RED

VEHICLE FRAME

SOCKET "A"

Step A. Disconnect the regulator-to-battery cable at the regulator.
Step B. Set the multimeter to read battery volts.
Step C. Ground the black probe and touch the red probe to socket A. (See Fig 12-1).
  1. If the meter reads 23-26 volts and:
    (a) In Check #5, the system was charging in the 27-29 volt range, but the batteries grow weak too often, do Test 6.
    (b) In Check #5, the system was not charging in the 27-29-volt range, do Test 2.
  2. If the meter does not read 23-26 volts, repair or replace the regulator-to-battery cable.

Test 2- Do the following steps to check the regulator's overvoltage relay:

NOTE: The M551 regulator contains an overvoltage relay that disconnects the generator from the batteries when an overvoltage condition exists. You can reset the relay by disconnecting and reconnecting the regulator-to-battery cable at the regulator.

Step A. Reconnect the regulator-to-battery cable.
Step B. Start the engine. Turn on the high beam lights. Idle the engine at 1000-1200 RPM. When the engine reaches normal operating temperature, do Step C.
Step C. Set the meter to read battery volts.
Step D. Measure the charging system output voltage at the batteries or at the slave receptacle.
  1. If the meter now reads 27-29 volts, the charging system checks OK.
  2. If the meter still does not read 27-29 volts, do Test 3.

Test 3- Do the following steps to test the generator and the generator-to-regulator cable for ohms:
Step A. Set the meter for ohms (RX1, X1). "Zero" the meter.
Step B. Disconnect the regulator-to-generator cable at the regulator.
Step C. Ground the black probe. Touch the red probe to Pin A, then to Pin B, Pin C and Pin D. Read the meter each time.

(Fig 12-2)

PIN "D"

PIN "B"

PIN "C"

PIN "A"

KEY

  1. If the meter reads 0 to 4 ohms each time, do Test 4.
  2. Be sure the connections to the generator are clean and tight. If the meter does not read 0 to 4 ohms each time, do Test 7 to check the generator and the cable.

Test 4- Do the following steps to test the generator for residual voltage:

Step A. Set the meter to read 10 volts.

Step B. Start the engine and run it at 1000-1200 RPM.

Step C. Ground the black probe. Touch the red probe to Pin A. Read the meter. (See Fig 12-2 for pin location.)

    1. If the meter reads between 1 and 4 volts, the generator is OK. Do Test 5 to see if adjusting the regulator voltage output rheostat will bring the voltage output into the 27-29 volt range.

    2. If the meter reads more than 4 volts, replace the generator.

    3. If the meter reads less than 1 volt or the needle deflects to the left of 0 volts (tries to go negative), do Steps D and E to polarize the generator.

Step D. Polarizing the Generator: Keep the engine off (master switch ON and generator ON-OFF switch ON for vehicles with a generator ON-OFF switch. Otherwise leave the master switch OFF). Use a heavy insulated jumper wire to jump from the middle battery cable (either terminal) to pin D on the regulator-to-generator cable. (See Fig 12-2 for pin location.) Repeat Steps B and C.

NOTE: If the reading is still bad (less than 1 volt or deflects left), repeat Step D. Do not attempt to polarize the generator more than twice. If the reading is bad after the second attempt, replace the generator. Then repeat Check 5, page 9.

NOTE: If local SOP prohibits adjustment at your level, replace the regulator. Then, repeat Check 5, page 9.

Test 5- Do the following steps to test or adjust the regulator voltage output:

Step A. Reconnect all cable assemblies.

Step B. Set multimeter up to read battery volts.

Step C. Start engine. Turn on high beam lights. Idle at 1000-1200 RPM (When engine reaches normal operating temperature, do Step D).

Step D. Leave engine at 1000-1200 RPM.

Step E. Ground the black probe, and touch the red probe to the positive terminal of the batteries or the slave receptacle.

    1. If the needle gives a steady reading in the 27-29 volt range, the charging system checks out OK.

    2. If the needle is not in the 27-29 volt range or is not steady, adjust the regulator output, Step F.

Step F. Remove the pipe plug from the regulator.

(Fig 12-4)

Step G. Turn the volt adjust screw with a small flat tip screwdriver until you get a steady 27-29 volt range reading. During hot weather, set the reading toward the low end of the range. During cold weather, set the reading toward the high end of the range.

    1. If you cannot get a steady reading in the 27-29 volt range, turn the engine off. Replace the regulator.

    2. If you get a steady reading in the 27-29 volt range, the charging system now checks out OK.

Test 6- Do the following steps to test the regulator's reverse current relay:

Step A. Hook up the regulator-to-battery cable.
Step B. Disconnect the regulator-to-generator cable at the regulator.
Step C. Set the meter to read battery volts.

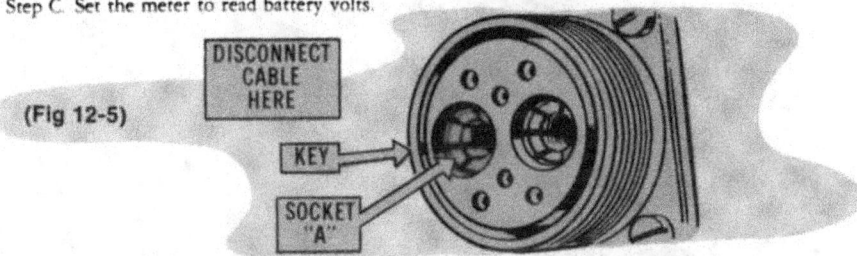

(Fig 12-5)

DISCONNECT CABLE HERE

KEY

SOCKET "A"

Step D. Ground the black probe and touch the red probe to Socket A of the regulator.
  1. If the meter reads 0 volts:
   (a) Check the batteries thoroughly (Checks #3 and #4, page 9) to be sure that there are no dead or shorted cells.
   (b) Check to see if the operators are operating accessories excessively without running the engine enough to keep the batteries up . . . or if they might be leaving electrical switches on after operating the vehicle.
  2. If the meter does not read 0 volts, replace the regulator, repeat Check 5, page 9.

Test 7- Do the following steps to test the regulator-to-generator cable for continuity and shorts.

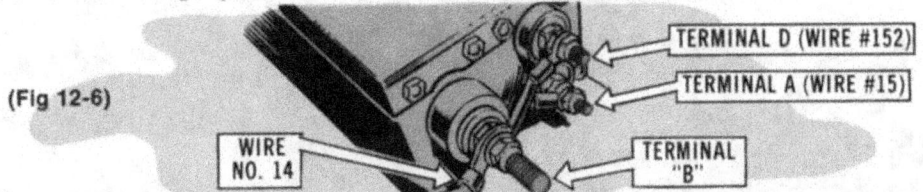

(Fig 12-6)

TERMINAL D (WIRE #152)
TERMINAL A (WIRE #15)
WIRE NO. 14
TERMINAL "B"

Step A. Remove the regulator-to-generator cable wires at the generator. Wire #14 goes to terminal B of the generator. Wire #15 goes to terminal A of the generator. Wire #152 goes to terminal D of the generator and wire #GND (braided) goes to terminal E in back of the generator.
Step B. Set the meter for ohms (RX1, X1). "Zero" the meter.
Step C. To check for continuity on the regulator-to-generator cable, touch either probe (black or red) to the pin and the other to the wire.

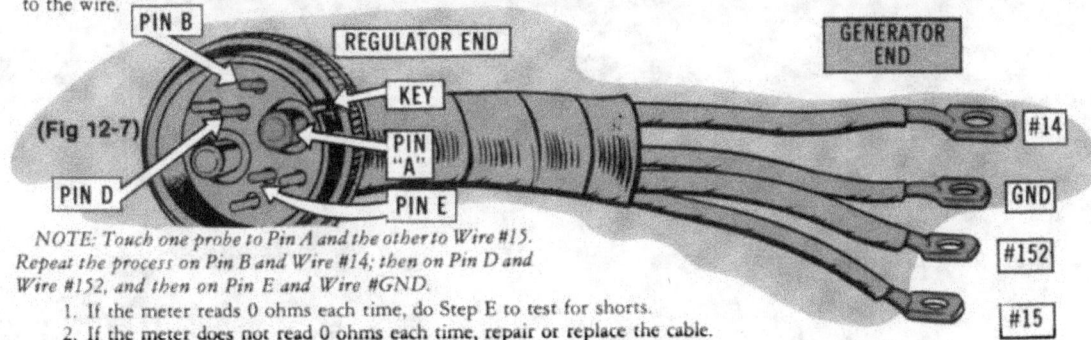

PIN B

REGULATOR END

GENERATOR END

(Fig 12-7)

KEY

PIN "A"

PIN D

PIN E

#14

GND

#152

#15

NOTE: *Touch one probe to Pin A and the other to Wire #15. Repeat the process on Pin B and Wire #14; then on Pin D and Wire #152, and then on Pin E and Wire #GND.*
  1. If the meter reads 0 ohms each time, do Step E to test for shorts.
  2. If the meter does not read 0 ohms each time, repair or replace the cable.

Step D. To check for shorts: (See Fig 12-7 for pin location). On the regulator end of the cable, touch the meter probes to the following pins and read the meter each time:
Black to pin A, red to pin B then to C and D.
Black to pin B, red to pin C then to D.
Black to pin C, red to pin D.
  1. If the meter needle moves, repair or replace the cable.
  2. If the meter needle does not move, the cable is good. Replace the defective generator.

END OF 300 AMP M551 CHARGING SYSTEM TESTS

☆ U. S. GOVERNMENT PRINTING OFFICE : 1977 O - 225-192

Forward comments on this pamphlet to:

Commander
US Army Maintenance Management Center
ATTN: DRXMD-TP
Lexington, KY 40511

By Order of the Secretary of the Army:

BERNARD W. ROGERS
General, United States Army
Chief of Staff

Official:

PAUL T. SMITH
Major General, United States Army
The Adjutant General

DISTRIBUTION: Regular Army, National Guard, Army Reserve.
Two copies each to all account holders on DA
Forms 12-37 and 12-38.

Forward comments on this pamphlet to:

Commander
US Army Maintenance Management Center
ATTN: DRXMD-TP
Lexington, KY 40511

By Order of the Secretary of the Army:

BERNARD W. ROGERS
General, United States Army
Chief of Staff

Official:

PAUL T. SMITH
Major General, United States Army
The Adjutant General

DISTRIBUTION: Regular Army, National Guard, Army Reserve.
Two copies each to all account holders on DA
Forms 12-37 and 12-38.